FREE Test Taking Tips DVD Offer

To help us better serve you, we have developed a Test Taking Tips DVD that we would like to give you for FREE. **This DVD covers world-class test taking tips that you can use to be even more successful when you are taking your test.**

All that we ask is that you email us your feedback about your study guide. Please let us know what you thought about it – whether that is good, bad or indifferent.

To get your **FREE Test Taking Tips DVD**, email freedvd@studyguideteam.com with "FREE DVD" in the subject line and the following information in the body of the email:

  a. The title of your study guide.

  b. Your product rating on a scale of 1-5, with 5 being the highest rating.

  c. Your feedback about the study guide. What did you think of it?

  d. Your full name and shipping address to send your free DVD.

If you have any questions or concerns, please don't hesitate to contact us at freedvd@studyguideteam.com.

Thanks again!

IB Biology Study Guide
IB Prep Book and Practice Test Questions for the Diploma Programme [Includes Detailed Answer Explanations]

Joshua Rueda

Written and edited by TPB Publishing.

TPB Publishing is not associated with or endorsed by any official testing organization. TPB Publishing is a publisher of unofficial educational products. All test and organization names are trademarks of their respective owners. Content in this book is included for utilitarian purposes only and does not constitute an endorsement by TPB Publishing of any particular point of view.

Interested in buying more than 10 copies of our product? Contact us about bulk discounts:
bulkorders@studyguideteam.com

ISBN 13: 9781637757055
ISBN 10: 1637757050

Table of Contents

Quick Overview

As you draw closer to taking your exam, effective preparation becomes more and more important. Thankfully, you have this study guide to help you get ready. Use this guide to help keep your studying on track and refer to it often.

This study guide contains several key sections that will help you be successful on your exam. The guide contains tips for what you should do the night before and the day of the test. Also included are test-taking tips. Knowing the right information is not always enough. Many well-prepared test takers struggle with exams. These tips will help equip you to accurately read, assess, and answer test questions.

A large part of the guide is devoted to showing you what content to expect on the exam and to helping you better understand that content. In this guide are practice test questions so that you can see how well you have grasped the content. Then, answer explanations are provided so that you can understand why you missed certain questions.

Don't try to cram the night before you take your exam. This is not a wise strategy for a few reasons. First, your retention of the information will be low. Your time would be better used by reviewing information you already know rather than trying to learn a lot of new information. Second, you will likely become stressed as you try to gain a large amount of knowledge in a short amount of time. Third, you will be depriving yourself of sleep. So be sure to go to bed at a reasonable time the night before. Being well-rested helps you focus and remain calm.

Be sure to eat a substantial breakfast the morning of the exam. If you are taking the exam in the afternoon, be sure to have a good lunch as well. Being hungry is distracting and can make it difficult to focus. You have hopefully spent lots of time preparing for the exam. Don't let an empty stomach get in the way of success!

When travelling to the testing center, leave earlier than needed. That way, you have a buffer in case you experience any delays. This will help you remain calm and will keep you from missing your appointment time at the testing center.

Be sure to pace yourself during the exam. Don't try to rush through the exam. There is no need to risk performing poorly on the exam just so you can leave the testing center early. Allow yourself to use all of the allotted time if needed.

Remain positive while taking the exam even if you feel like you are performing poorly. Thinking about the content you should have mastered will not help you perform better on the exam.

Once the exam is complete, take some time to relax. Even if you feel that you need to take the exam again, you will be well served by some down time before you begin studying again. It's often easier to convince yourself to study if you know that it will come with a reward!

Test-Taking Strategies

1. Predicting the Answer

When you feel confident in your preparation for a multiple-choice test, try predicting the answer before reading the answer choices. This is especially useful on questions that test objective factual knowledge. By predicting the answer before reading the available choices, you eliminate the possibility that you will be distracted or led astray by an incorrect answer choice. You will feel more confident in your selection if you read the question, predict the answer, and then find your prediction among the answer choices. After using this strategy, be sure to still read all of the answer choices carefully and completely. If you feel unprepared, you should not attempt to predict the answers. This would be a waste of time and an opportunity for your mind to wander in the wrong direction.

2. Reading the Whole Question

Too often, test takers scan a multiple-choice question, recognize a few familiar words, and immediately jump to the answer choices. Test authors are aware of this common impatience, and they will sometimes prey upon it. For instance, a test author might subtly turn the question into a negative, or he or she might redirect the focus of the question right at the end. The only way to avoid falling into these traps is to read the entirety of the question carefully before reading the answer choices.

3. Looking for Wrong Answers

Long and complicated multiple-choice questions can be intimidating. One way to simplify a difficult multiple-choice question is to eliminate all of the answer choices that are clearly wrong. In most sets of answers, there will be at least one selection that can be dismissed right away. If the test is administered on paper, the test taker could draw a line through it to indicate that it may be ignored; otherwise, the test taker will have to perform this operation mentally or on scratch paper. In either case, once the obviously incorrect answers have been eliminated, the remaining choices may be considered. Sometimes identifying the clearly wrong answers will give the test taker some information about the correct answer. For instance, if one of the remaining answer choices is a direct opposite of one of the eliminated answer choices, it may well be the correct answer. The opposite of obviously wrong is obviously right! Of course, this is not always the case. Some answers are obviously incorrect simply because they are irrelevant to the question being asked. Still, identifying and eliminating some incorrect answer choices is a good way to simplify a multiple-choice question.

4. Don't Overanalyze

Anxious test takers often overanalyze questions. When you are nervous, your brain will often run wild, causing you to make associations and discover clues that don't actually exist. If you feel that this may be a problem for you, do whatever you can to slow down during the test. Try taking a deep breath or counting to ten. As you read and consider the question, restrict yourself to the particular words used by the author. Avoid thought tangents about what the author *really* meant, or what he or she was *trying* to say. The only things that matter on a multiple-choice test are the words that are actually in the question. You must avoid reading too much into a multiple-choice question, or supposing that the writer meant something other than what he or she wrote.

5. No Need for Panic

It is wise to learn as many strategies as possible before taking a multiple-choice test, but it is likely that you will come across a few questions for which you simply don't know the answer. In this situation, avoid panicking. Because most multiple-choice tests include dozens of questions, the relative value of a single wrong answer is small. As much as possible, you should compartmentalize each question on a multiple-choice test. In other words, you should not allow your feelings about one question to affect your success on the others. When you find a question that you either don't understand or don't know how to answer, just take a deep breath and do your best. Read the entire question slowly and carefully. Try rephrasing the question a couple of different ways. Then, read all of the answer choices carefully. After eliminating obviously wrong answers, make a selection and move on to the next question.

6. Confusing Answer Choices

When working on a difficult multiple-choice question, there may be a tendency to focus on the answer choices that are the easiest to understand. Many people, whether consciously or not, gravitate to the answer choices that require the least concentration, knowledge, and memory. This is a mistake. When you come across an answer choice that is confusing, you should give it extra attention. A question might be confusing because you do not know the subject matter to which it refers. If this is the case, don't eliminate the answer before you have affirmatively settled on another. When you come across an answer choice of this type, set it aside as you look at the remaining choices. If you can confidently assert that one of the other choices is correct, you can leave the confusing answer aside. Otherwise, you will need to take a moment to try to better understand the confusing answer choice. Rephrasing is one way to tease out the sense of a confusing answer choice.

7. Your First Instinct

Many people struggle with multiple-choice tests because they overthink the questions. If you have studied sufficiently for the test, you should be prepared to trust your first instinct once you have carefully and completely read the question and all of the answer choices. There is a great deal of research suggesting that the mind can come to the correct conclusion very quickly once it has obtained all of the relevant information. At times, it may seem to you as if your intuition is working faster even than your reasoning mind. This may in fact be true. The knowledge you obtain while studying may be retrieved from your subconscious before you have a chance to work out the associations that support it. Verify your instinct by working out the reasons that it should be trusted.

8. Key Words

Many test takers struggle with multiple-choice questions because they have poor reading comprehension skills. Quickly reading and understanding a multiple-choice question requires a mixture of skill and experience. To help with this, try jotting down a few key words and phrases on a piece of scrap paper. Doing this concentrates the process of reading and forces the mind to weigh the relative importance of the question's parts. In selecting words and phrases to write down, the test taker thinks about the question more deeply and carefully. This is especially true for multiple-choice questions that are preceded by a long prompt.

9. Subtle Negatives

One of the oldest tricks in the multiple-choice test writer's book is to subtly reverse the meaning of a question with a word like *not* or *except*. If you are not paying attention to each word in the question, you can easily be led astray by this trick. For instance, a common question format is, "Which of the following is…?" Obviously, if the question instead is, "Which of the following is not…?," then the answer will be quite different. Even worse, the test makers are aware of the potential for this mistake and will include one answer choice that would be correct if the question were not negated or reversed. A test taker who misses the reversal will find what he or she believes to be a correct answer and will be so confident that he or she will fail to reread the question and discover the original error. The only way to avoid this is to practice a wide variety of multiple-choice questions and to pay close attention to each and every word.

10. Reading Every Answer Choice

It may seem obvious, but you should always read every one of the answer choices! Too many test takers fall into the habit of scanning the question and assuming that they understand the question because they recognize a few key words. From there, they pick the first answer choice that answers the question they believe they have read. Test takers who read all of the answer choices might discover that one of the latter answer choices is actually *more* correct. Moreover, reading all of the answer choices can remind you of facts related to the question that can help you arrive at the correct answer. Sometimes, a misstatement or incorrect detail in one of the latter answer choices will trigger your memory of the subject and will enable you to find the right answer. Failing to read all of the answer choices is like not reading all of the items on a restaurant menu: you might miss out on the perfect choice.

11. Spot the Hedges

One of the keys to success on multiple-choice tests is paying close attention to every word. This is never truer than with words like almost, most, some, and sometimes. These words are called "hedges" because they indicate that a statement is not totally true or not true in every place and time. An absolute statement will contain no hedges, but in many subjects, the answers are not always straightforward or absolute. There are always exceptions to the rules in these subjects. For this reason, you should favor those multiple-choice questions that contain hedging language. The presence of qualifying words indicates that the author is taking special care with their words, which is certainly important when composing the right answer. After all, there are many ways to be wrong, but there is only one way to be right! For this reason, it is wise to avoid answers that are absolute when taking a multiple-choice test. An absolute answer is one that says things are either all one way or all another. They often include words like *every*, *always*, *best*, and *never*. If you are taking a multiple-choice test in a subject that doesn't lend itself to absolute answers, be on your guard if you see any of these words.

12. Long Answers

In many subject areas, the answers are not simple. As already mentioned, the right answer often requires hedges. Another common feature of the answers to a complex or subjective question are qualifying clauses, which are groups of words that subtly modify the meaning of the sentence. If the question or answer choice describes a rule to which there are exceptions or the subject matter is complicated, ambiguous, or confusing, the correct answer will require many words in order to be expressed clearly and accurately. In essence, you should not be deterred by answer choices that seem excessively long. Oftentimes, the author of the text will not be able to write the correct answer without

offering some qualifications and modifications. Your job is to read the answer choices thoroughly and completely and to select the one that most accurately and precisely answers the question.

13. Restating to Understand

Sometimes, a question on a multiple-choice test is difficult not because of what it asks but because of how it is written. If this is the case, restate the question or answer choice in different words. This process serves a couple of important purposes. First, it forces you to concentrate on the core of the question. In order to rephrase the question accurately, you have to understand it well. Rephrasing the question will concentrate your mind on the key words and ideas. Second, it will present the information to your mind in a fresh way. This process may trigger your memory and render some useful scrap of information picked up while studying.

14. True Statements

Sometimes an answer choice will be true in itself, but it does not answer the question. This is one of the main reasons why it is essential to read the question carefully and completely before proceeding to the answer choices. Too often, test takers skip ahead to the answer choices and look for true statements. Having found one of these, they are content to select it without reference to the question above. Obviously, this provides an easy way for test makers to play tricks. The savvy test taker will always read the entire question before turning to the answer choices. Then, having settled on a correct answer choice, he or she will refer to the original question and ensure that the selected answer is relevant. The mistake of choosing a correct-but-irrelevant answer choice is especially common on questions related to specific pieces of objective knowledge. A prepared test taker will have a wealth of factual knowledge at their disposal, and should not be careless in its application.

15. No Patterns

One of the more dangerous ideas that circulates about multiple-choice tests is that the correct answers tend to fall into patterns. These erroneous ideas range from a belief that B and C are the most common right answers, to the idea that an unprepared test-taker should answer "A-B-A-C-A-D-A-B-A." It cannot be emphasized enough that pattern-seeking of this type is exactly the WRONG way to approach a multiple-choice test. To begin with, it is highly unlikely that the test maker will plot the correct answers according to some predetermined pattern. The questions are scrambled and delivered in a random order. Furthermore, even if the test maker was following a pattern in the assignation of correct answers, there is no reason why the test taker would know which pattern he or she was using. Any attempt to discern a pattern in the answer choices is a waste of time and a distraction from the real work of taking the test. A test taker would be much better served by extra preparation before the test than by reliance on a pattern in the answers.

FREE DVD OFFER

Don't forget that doing well on your exam includes both understanding the test content and understanding how to use what you know to do well on the test. We offer a completely FREE Test Taking Tips DVD that covers world class test taking tips that you can use to be even more successful when you are taking your test.

All that we ask is that you email us your feedback about your study guide. To get your **FREE Test Taking Tips DVD**, email freedvd@studyguideteam.com with "FREE DVD" in the subject line and the following information in the body of the email:

- The title of your study guide.
- Your product rating on a scale of 1-5, with 5 being the highest rating.
- Your feedback about the study guide. What did you think of it?
- Your full name and shipping address to send your free DVD.

Introduction

Function of the Test

The International Baccalaureate (IB) is a curriculum and exam used worldwide in governments, international organizations, and schools to assess young people in their development of knowledge and compassion. The IB is composed of four programs: Primary Years, Middle Years, Diploma, and Career-Related. This guide will focus on the Diploma program, specifically Biology. This program is for students sixteen to nineteen years old who are at schools that authorize the Diploma program and is geared toward helping students get prepared for university level.

Students may send their IB Biology scores to six universities for free, as their transcripts will help universities determine whether they are a good fit for the school. The IB Diploma program (DP) has been used for advanced placement, scholarships, as well as other admissions benefits. On average, around 80 percent of Diploma program students are awarded the diploma per examination session. Only students enrolled in an authorized IB World School can earn the IB diploma.

Test Administration

The IB Diploma exam times are divided up by afternoon sessions and morning sessions twice a year, once at the end of April and throughout the month of May, and once at the end of October and throughout the month of November. If test takers receive a grade and schools believe that the grade is undeserving, the school can request a re-mark, also known as a re-test. However, test takers cannot request a re-mark themselves. A test taker can retake a subject at a future examination session, but restrictions apply, so they must contact their DP coordinator.

Test Format

The core components of the IB DP Biology program features the following:

- Cell biology
- Molecular biology
- Genetics
- Ecology
- Evolution and biodiversity
- Human physiology

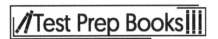

The biology assessment has an external assessment and an internal assessment. The following two tables depict the two types of assessments:

External Assessment (4.5 hours long, 80% of grade)			
Type	Format	Time	Percentage
Paper 1	40 multiple choice	1 hour	20%
Paper 2	Data based, short answer and extended response	2.25 hours	36%
Paper 3	Data based, short answer and extended response	1.25 hours	24%

Internal Assessment (10 hours long, 20% of grade)			
Type	Format	Time	Percentage
Individual Investigation	Investigation and write-up 6 to 12 pages	10 hours	20%

Scoring

For the High-Level IB Biology exam in November 2019, the mean grade was a 4.61 out of 7. A number of 2,595 students sat for this session. The IB uses a 1 to 7 range, with 7 being the highest score a student can receive. A score of 4 is considered passing.

Study Prep Plan for the IB Biology Exam

1   **Schedule -** Use one of our study schedules below or come up with one of your own.

2   **Relax -** Test anxiety can hurt even the best students. There are many ways to reduce stress. Find the one that works best for you.

3   **Execute -** Once you have a good plan in place, be sure to stick to it.

One Week Study Schedule

Day 1	Cell Biology
Day 2	Molecular Biology
Day 3	Genetics
Day 4	Ecology
Day 5	Human Physiology
Day 6	Practice Test
Day 7	Take Your Exam!

Two Week Study Schedule

Day 1	Cell Biology	Day 8	Law of Addition (Mutually Exclusive Results)
Day 2	Membrane Structure	Day 9	Ecology
Day 3	Cell Division	Day 10	Evolution and Biodiversity
Day 4	Molecular Biology	Day 11	Human Physiology
Day 5	DNA and RNA	Day 12	Neurons and Synapses
Day 6	Photosynthesis	Day 13	Practice Test
Day 7	Genetics	Day 14	Take Your Exam!

One Month Study Schedule					
Day 1	Introduction to Cells	Day 11	Cell Respiration	Day 21	Evolution and Biodiversity
Day 2	Membrane Structure	Day 12	Photosynthesis	Day 22	The Effect of Phenotypic Variations...
Day 3	Membrane Transport	Day 13	Practice Questions	Day 23	Practice Questions
Day 4	The Origin of Cells	Day 14	Genes	Day 24	Digestion and Absorption
Day 5	Cell Division	Day 15	Meiosis	Day 25	Defense Against Infectious Disease
Day 6	Practice Questions	Day 16	Genetic Crosses	Day 26	Neurons and Synapses
Day 7	Molecules to Metabolism	Day 17	Practice Questions	Day 27	Hormones, Homeostasis, and...
Day 8	Carbohydrates and Lipids	Day 18	Species, Communities, and...	Day 28	Practice Questions
Day 9	DNA and RNA	Day 19	Energy Flow	Day 29	Practice Test
Day 10	DNA Replication, Transcription, and Translation	Day 20	Practice Questions	Day 30	Take Your Exam!

Cell Biology

Introduction to Cells

Cell Structure and Function

The **cell** is the main functional and structural component of all living organisms. Robert Hooke, an English scientist, coined the term "cell" in 1665. Hooke's discovery laid the groundwork for the **cell theory**, which is composed of three principles:

1. All organisms are composed of cells.
2. All existing cells are created from other living cells.
3. The cell is the most fundamental unit of life.

Organisms can be **unicellular** (composed of one cell) or **multicellular** (composed of many cells). All cells are bounded by a cell membrane, filled with cytoplasm of some sort, and are coded for by a genetic sequence.

The cell membrane separates a cell's internal and external environments. It is a **selectively permeable** membrane, which usually only allows the passage of certain molecules by diffusion. Phospholipids and proteins are crucial components of all cell membranes. The **cytoplasm** is the cell's internal environment and is aqueous, or water-based. The **genome** represents the genetic material inside the cell that is passed on from generation to generation.

The Effects of Surface Area-to-Volume Ratios

Organisms require certain reactants to sustain life and have adaptations that maximize access to these reactants. For example, the small intestine contains villi and microvilli, which are millions of small projections that absorb nutrients and deliver them to the circulatory system. The villi and microvilli provide extra surface area to maximize absorption. This is similar to a towel, in that the fluffier the towel the more effective it is at absorption. Root hairs serve the same purpose in plants.

The larger an organism is, the more it needs such structures to maintain adequate levels of material exchange. Single-celled organisms typically remain small to keep a large surface area to volume (SA:V) ratio that will maximize the exchange of materials with the surroundings.

To calculate surface area to volume ratio, simply divide the volume (V) from the surface area (SA):

$$SA/V = Ratio$$

In order to do so, one must consider the shape of the organism for which they are calculating. Most organisms are not just one simple shape, and so calculations for a whole organism will not need to be performed. However, simple shapes, such as cells or bacteria, have a relatively simple calculation.

Here are the common shapes associated with the different kinds of cells, and their corresponding volume and surface area equations:

Shape	Surface Area	Volume
Cube (cuboidal and squamous)	$6a^2$	a^3
Sphere (spherical)	$4\pi r^2$ $\pi = \sim 3.14$	$(4/3)\pi r^3$
Cylinder (columnar)	$2\pi r^2 + 2\pi rh$	$\pi r^2 h$
Rectangular Prism	$2\,(wl + hw + hl)$	$L \times W \times H$

Cell Differentiation

Cell differentiation refers to the process of a cell transforming into another type of cell. It most commonly involves a less specialized cell transforming into a more specialized cell.

The human body contains a vast array of cells which undergo division and differentiation to compose each unique human being. The trillions of cells composing the human body are derived from one cell, a

fertilized egg called a **zygote**. The zygote not only divides, but also differentiates into cells that perform specific tasks.

Genes control the process of cell differentiation during human development. The zygote divides through mitosis into a blastula and then into a gastrula. At this stage, the three embryonic germ layers (endoderm, mesoderm, and ectoderm) are formed. Most of the human body systems develop from one or more of the embryonic germ layers. For example, the digestive system develops from the **endoderm**, or innermost germ layer; the cardiovascular system develops from the **mesoderm**, or middle germ layer; and the nervous system develops from the **ectoderm**, or outer germ layer.

Ultrastructure of Cells

Prokaryotes and Eukaryotes

Prokaryotic cells are much smaller than eukaryotic cells. The vast majority of prokaryotes are unicellular, while the majority of eukaryotes are multicellular. Prokaryotic cells have no nucleus, and their genome is found in an area known as the **nucleoid**. They also do not have membrane-bound organelles, which are "little organs" that perform specific functions within a cell.

Eukaryotic cells have a proper nucleus, which contains the genome. They also have numerous membrane-bound organelles such as lysosomes, endoplasmic reticula (rough and smooth), Golgi complexes, and mitochondria.

The majority of prokaryotic cells have cell walls, while most eukaryotic cells do not have cell walls. The DNA of prokaryotic cells is contained in a single circular chromosome, while the DNA of eukaryotic cells is contained in multiple linear chromosomes. Prokaryotic cells divide using binary fission, while eukaryotic cells divide using mitosis. Examples of prokaryotes are bacteria and archaea, while examples of eukaryotes are animals and plants.

Structure and Function of Prokaryotic Organelles

Prokaryotic cells are usually structurally less complex than eukaryotic cells because they lack membrane-bound organelles. While they do contain some of the same structures as eukaryotic cells—including ribosomes, cytoplasm, and plasma membranes—they have several unique structures as well including the following:

- **Chromosome:** Located in the nucleotide region of the cell, the bacterial chromosome is a single loop of DNA that carries the genes for the cell's protein synthesis.

- **Plasmids:** Some bacterial cells have plasmids, which are small extra-chromosomal rings of DNA that contain genes that code for resistance to various antibiotics. Through transposons, genetic information can be moved between the chromosome and plasmids.

- **Cell wall:** Like plant cells, most bacteria have a cell wall on the outer surface of their cell membrane. It usually contains peptidoglycan in either a single or double layer. **Gram-positive** bacteria have a single layer of peptidoglycan in their cell walls, while **gram-negative** bacteria have a double layer.

- **Flagella:** Flagella provide motility to the cell but differ in structure from those of eukaryotic cells because they have a hollow helical confirmation anchored into the cell membrane. The power

for their rotation is provided by a proton pump in the cell membrane. Prokaryotic cells may have one, multiple, or no flagella; of course, those that contain more are more mobile.

- **Spores:** There are a few species of prokaryotes that can create spores during unfavorable environmental conditions, which enables them to survive for several years in spore form and then germinate into the vegetative cell form when conditions improve.

- **Capsule** and **slime layers:** Some bacterial cells have a sticky layer of sugars and proteins on their outer surface, which enables the cell to attach to various surfaces.

- **Pili:** Some prokaryotes have pili, which are tiny proteins covering the cell's surface. Like slime layers, pili assist in the attachment of the cell to other surfaces.

Structure and Function of Eukaryotic Cellular Organelles

Organelles are specialized structures that perform specific tasks in a cell. The term literally means "little organs." Most organelles are membrane bound and serve as sites for the production or degradation of chemicals. The following are organelles found in eukaryotic cells:

Nucleus

A cell's **nucleus** contains genetic information in the form of DNA. The nucleus is surrounded by the nuclear envelope. A single nucleus is the defining characteristic of eukaryotic cells. The nucleus is also the most important organelle of the cell. It contains the nucleolus, which manufactures ribosomes (another organelle) that are crucial in **protein synthesis** (also called **gene expression**).

Mitochondria

The **mitochondrion** is the primary site of respiration and adenosine triphosphate (ATP) synthesis inside the cell. Mitochondria have two lipid bilayers that create the **intermembrane space**, which is the space between the two membranes, and the **matrix**, which is the space inside the inner membrane. While the outer membrane is smooth, the inner membrane is folded and forms cristae. The outer membrane is permeable to small molecules. The cristae contain many proteins involved in ATP synthesis. In the mitochondria, the products of glycolysis are further oxidized during the citric-acid cycle and the electron-transport chain. The two-layer structure of the mitochondria allows for the buildup of H+ ions produced during the electron-transport chain in the intermembrane space, which creates a proton

gradient and an energy potential. This gradient drives the formation of ATP. Mitochondria have their own DNA and are capable of replication.

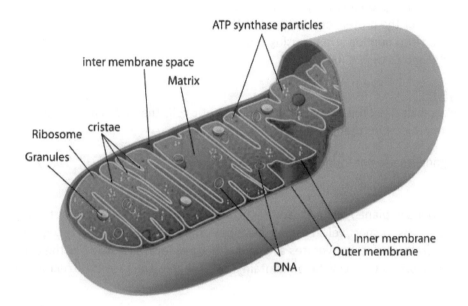

Rough Endoplasmic Reticulum

The **rough endoplasmic reticulum (RER)** is composed of linked membranous sacs called cisternae with ribosomes attached to their external surfaces. The RER is responsible for the production of proteins that will eventually get shipped out of the cell.

Smooth Endoplasmic Reticulum

The **smooth endoplasmic reticulum (SER)** is composed of linked membranous sacs called cisternae without ribosomes, which distinguishes it from the RER. The SER's main function is the production of carbohydrates and lipids which can be created expressly for the cell or to modify the proteins from the RER that will eventually get shipped out of the cell.

Golgi Apparatus

The **Golgi apparatus** is the site where proteins are modified; it is involved in the transport of proteins, lipids, and carbohydrates within the cell. The Golgi apparatus is made up of flat layers of membranes called cisternae. Material is transported in transfer vesicles from the ER to the cis region of the Golgi apparatus. From there, the material moves through the medial region, where it is sometimes modified, and then leaves through the trans region of the ER in a secretory vesicle.

Lysosomes

Lysosomes are specialized vesicles that contain enzymes capable of digesting food, surplus organelles, and foreign invaders such as bacteria and viruses. They often destroy dead cells in order to recycle cellular components. Lysosomes are found only in animal cells.

Secretory Vesicles

Secretory vesicles transport and deliver molecules into or out of the cell via the cell membrane. **Endocytosis** refers to the movement of molecules into a cell via secretory vesicles. **Exocytosis** refers to the movement of molecules out of a cell via secretory vesicles.

Ribosomes

Ribosomes are made up of ribosomal RNA molecules and a variety of proteins, and they are the structures that synthesize proteins. They consist of two subunits: small and large. The ribosomes use mRNA as a template for the protein and they use tRNA to bring amino acids to the ribosomes, where they are synthesized into peptide strands using the genetic code provided by the messenger RNA. Most ribosomes are attached to the ER membrane.

Cilia and Flagella

Cilia are specialized hair-like projections on some eukaryotic cells that aid in movement, while **flagella** are long, whip-like projections that are used in the same capacity.

The following organelles are *not* found in animal cells:

Cell Walls

Cell walls can be found in plants, bacteria, and fungi, and are made of cellulose, peptidoglycan, and lignin in these organisms, respectively. Each of these materials is a type of sugar recognized as a structural carbohydrate. The carbohydrates are rigid structures located outside of the cell membrane. Cell walls function to protect the cell, help maintain a cell's shape, and provide structural support.

Chloroplasts

Chloroplasts are organelles found primarily in plants and are the site of photosynthesis. They have a double membrane and also contain membrane-bound **thylakoids**, or discs, that are organized into **grana**, or stacks. Chlorophyll is present in the thylakoids, and the light stage of photosynthesis—which includes the production of ATP and $NADPH_2$—occurs there. Chlorophyll is green and traps the light energy necessary for photosynthesis. Chloroplasts also contain **stroma**—which are the site of the dark reaction stage of photosynthesis—during which sugar is made. The membrane structure of chloroplasts allows for the compartmentalization of the light and dark stages of photosynthesis.

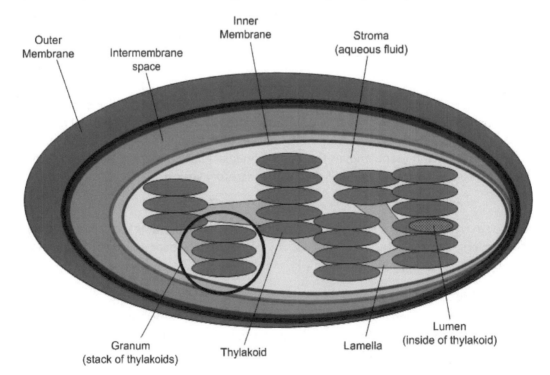

Vacuoles

Vacuoles are membrane-bound organelles primarily found in plant and fungi cells, but also in some animal cells. Vacuoles are filled with water and some enzymes and are important for intracellular digestion and waste removal. The membrane-bound nature of the vacuole allows for the storage of harmful material and poisonous substances. The pressure from the water inside the vacuole also contributes to the structure of plant cells.

Here is an illustration of the cell:

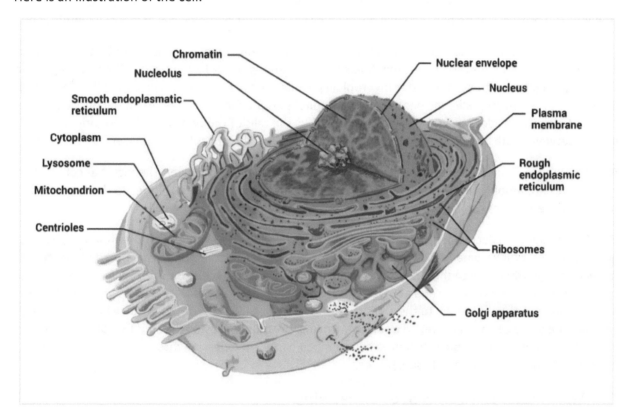

Nuclear Parts of a Cell

Nucleus (plural nuclei): Houses a cell's genetic material, deoxyribonucleic acid (DNA), which is used to form chromosomes. A single nucleus is the defining characteristic of eukaryotic cells. The nucleus of a cell controls gene expression. It ensures genetic material is transmitted from one generation to the next.

Chromosomes: Complex thread-like arrangements composed of DNA that is found in a cell's nucleus. Humans have 23 pairs of chromosomes, for a total of 46.

Chromatin: An aggregate of genetic material, consisting of DNA and proteins, that forms chromosomes during cell division.

Nucleolus (plural nucleoli): The largest component of the nucleus of a eukaryotic cell. With no membrane, the primary function of the nucleolus is the production of ribosomes, which are crucial to the synthesis of proteins.

Membrane Structure

Cell Membranes

Cell membranes enclose the cell's cytoplasm, separating the intracellular environment from the extracellular environment. They are selectively permeable, which enables them to control molecular traffic entering and exiting cells. Cell membranes are made of a double layer of phospholipids studded with proteins. Cholesterol is also dispersed in the phospholipid bilayer of cell membranes to provide stability. The proteins in the phospholipid bilayer aid the transport of molecules across cell membranes.

Scientists use the term "fluid mosaic model" to refer to the arrangement of phospholipids and proteins in cell membranes. In this model, phospholipids have a head region and a tail region. The head region of the phospholipids is attracted to water (hydrophilic), while the tail region is repelled by it (hydrophobic). Because they are hydrophilic, the heads of the phospholipids are facing the water, on the outside of the cell and lining the inside. Because they are hydrophobic, the tails of the phospholipids are oriented inward between both head regions. This orientation constructs the phospholipid bilayer.

As mentioned, cell membranes have the distinct trait of selective permeability. The fact that cell membranes are **amphiphilic** (having hydrophilic and hydrophobic zones) contributes to this trait. As a result, cell membranes can regulate the flow of molecules into and out of the cell.

Factors relating to molecules, such as size, polarity, and solubility, determine their likelihood of passage across cell membranes. Small molecules can diffuse easily across cell membranes compared to large molecules. **Polarity** refers to the charge present in a molecule. **Polar molecules** have regions, or poles, of positive and negative charge and are water-soluble, while **non-polar molecules** have no charge and are fat-soluble. **Solubility** refers to the ability of a substance, called a **solute**, to dissolve in a solvent. A **soluble** substance can be dissolved in a solvent, while an **insoluble** substance cannot be dissolved in a solvent. Non-polar, fat-soluble substances have a much easier time passing through cell membranes compared to polar, water-soluble substances.

Cell Membranes Have Selective Permeability

Water is polar because of the bent shape of the molecule—one side of the molecule (the oxygen side) has a negative charge, and the other side (the hydrogen side) has a positive charge. Oxygen, though more negative than hydrogen, has a more positively charged nucleus, causing it to pull the two hydrogen molecules' electrons closer to itself. This results in a partial positive charge of the hydrogen side and makes oxygen more electronegative (see the following figure). The molecule is bent because the electrons pulled from hydrogen are not bonded to any other atoms, and are termed "lone" electrons. Further, the two hydrogen atoms on the opposite side of the oxygen are repelled by their own positive forces, and the lone electrons are also repelled by their similar charges. Therefore, they stay as far away from each other as possible while still being held in place by oxygen.

Polar substances dissolve other polar substances. For example, when table salt (NaCl) is placed in water, the molecule is split into Na^+ and Cl^- because Na^+ is attracted to $O2^-$, and Cl^- is attracted to the H^+, essentially "tearing" the molecules apart. Polar substances are said to be hydrophilic (water-loving), and non-polar substance are said to be hydrophobic (water-fearing).

H₂O bond

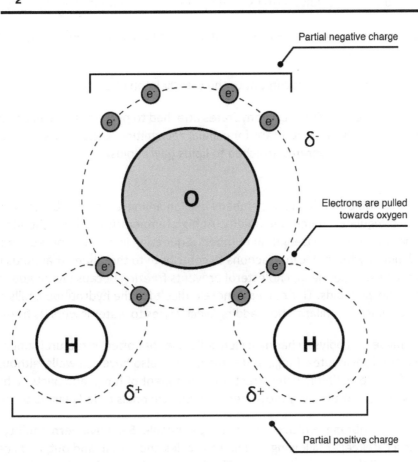

The cell membrane in bacteria and eukaryotes is a phospholipid bilayer that separates the extracellular and intracellular microsystems. It is a dynamic, fluid, and heterogeneous entity commonly described as a fluid mosaic that contains the following structures:

Phospholipid Bilayer

The bilayer is composed of polar hydrophilic phosphate "heads" on the exterior and two hydrophobic fatty acid "tails" on the interior. The phospholipid-charged regions are referred to as the heads because they face the extracellular and intracellular sides. The hydrophobic non-polar tails meet in the middle. There is lateral movement between adjacent phospholipids quite often. Flip-flopping of phospholipids transversely across the membrane from the intracellular to the extracellular side (or vice-versa) is very rare.

Integral and Peripheral Proteins

Proteins may be completely embedded in the membrane if they are integral, or just attached to one side if they are peripheral proteins. Their placement in the membrane depends on their folding and the polarity or non-polarity of their exposed regions. These proteins can have many different functions.

- Integral proteins can act as channels for movement.

- Enzymes are biological catalysts that speed up chemical reactions.

- Membrane receptors can receive ligands (smaller molecules) and initiate signal transduction cascades.

- Proteins are important for attachment to the cytoskeleton or extracellular matrix.

- Glycoproteins are extracellular carbohydrates attached to proteins and can act as a cell identifier/marker, which is important for cellular recognition. Other markers can be free carbohydrates or carbohydrates attached to lipids (glycolipids).

Cholesterol

Cholesterol is a hydrophobic steroid lipid. It embeds itself in animal cells' lipid bilayers between hydrophobic tails and regulates membrane fluidity. At high temperatures, the embedded cholesterol prevents melting because there is strength in numbers. Squeezing in an extra non-polar molecule provides for many more hydrophobic interactions to contribute to the "glue" that holds it together. In cold temperatures, on the other hand, cholesterol prevents freezing because its ringed structure interrupts adjacent phospholipids. This makes it more difficult for the hydrophobic tails to line up perfectly and freeze, which is similar to how adding electrolytes to water lowers its freezing point.

Plant cell walls are made of a polysaccharide called **cellulose** that offers structural support and protection to the plant. Prokaryotes, fungi, and some protists also have cell walls, although they are not necessarily made of cellulose. Chitin is the structural component of fungi, and bacteria have a cell wall made of peptidoglycan. Cell walls allow movement through channels called plasmodesmata.

All cells contain a cell membrane, which is selectively permeable. **Selective permeability** means essentially that it is a gatekeeper, allowing certain molecules and ions in and out, and keeping unwanted ones at bay, at least until they are ready for use. This is achieved through active and passive transport, actively allowing molecules and ions through the opening and closing of cell membranes embedded within the phospholipid bilayer (using energy), or passively via a concentration gradient.

The cell membrane, or plasma membrane, has selective permeability with regard to size, charge, and solubility. With regard to molecule size, the cell membrane allows only small molecules to diffuse through it. Oxygen and water molecules are small and typically can pass through the cell membrane. The charge of the ions on the cell's surface also either attracts or repels ions. Ions with like charges are repelled, and ions with opposite charges are attracted to the cell's surface. Molecules that are soluble in phospholipids can usually pass through the cell membrane. Many molecules are not able to diffuse the cell membrane, and, if needed, those molecules must be moved through by active transport and vesicles.

Membrane Transport

Passive Transport

Passive transport is the movement of substances from high concentration to low concentration without the use of energy. This occurs because of the universe's tendency to achieve a state of equilibrium, or balance. For example, when food coloring is dropped into a cup of water, it spreads to the rest of the water until equilibrium is reached and there is a homologous solution. Passive transport comes in the following three varieties:

- **Diffusion**: When particles move from high concentration to low concentration, like when hot chocolate powder dissolves in water to form a tasty treat.

- **Osmosis**: When water moves from high concentration to low concentration via a permeable membrane. Water moves to the higher concentration of solutes in order to achieve a more equal water to solute ratio.

- **Facilitated diffusion**: When particles diffuse through a channel protein due to the membrane's selective permeability

Water is a molecule that travels via facilitated diffusion through proteins called **aquaporins**. Water movement is determined by its environment, of which there are three types:

- Isotonic environments are achieved when there is a dynamic equilibrium between two solutions. Water will move in and out at equal rates.

- Hypotonic solutions are ones that have a lower solute concentration than the solution they are being compared to. The solution will be low in solute and high in water, so water will move out to achieve equilibrium.

- Hypertonic solutions are ones that have a higher solute concentration than the one they are being compared to. They will be high in solute and low in water, so water will move in to achieve equilibrium.

Animal cells in a hypotonic environment are subject to a net water movement into the cell, so the cell will swell and possibly burst if equilibrium is never achieved. Conversely, hypertonic solutions will cause the cell to shrink as water moves out.

Plant cells are also affected by their water concentrations. The pressure of their cell wall against the cell membrane is called **turgor pressure**. In hypotonic environments, the cell will swell and the cell wall will press on the cell membrane, resulting in a turgid, or firm, cell. This is healthiest for a plant because it provides the support to hold it upright. Isotonic environments mean that water will diffuse in and out of the plants cells at equal rates. In a hypertonic environment, the plant cell will undergo a process called **plasmolysis**, where the cell membrane shrinks and separates from the cell wall, resulting in a wilted plant due to lack of turgor pressure.

Protists such as paramecium that live in hypotonic environments will constantly have water moving inward and will never achieve equilibrium. They have a contractile vacuole that actually pumps out extra water using energy.

Water potential (Ψ) always indicates the direction of net water flow. It is determined by the sum of the pressure potential (Ψ_p) and solute potential (Ψ_s).

$$\Psi = \Psi_s + \Psi_p$$

Pressure potential is the force a plant cell wall exerts on its cell membrane (turgor pressure). Solute potential (also called osmotic potential) is defined by the pressure needed to be added to a solution in order to prevent the influx of water into the other solution. It is influenced by concentration, by molarity of the solution, by ionization of the solute, and by temperature of the system. Solute potential decreases as solute is added to a solution, thus the negative sign in the equation below. Distilled water has a solute potential of zero.

$$\Psi_s = -iCRT$$

The equation above shows the relationship between the variables. R is not a variable—it is the pressure constant equal to $.0831\ L \times bar \div mole \times K$. A bar is a very large unit of pressure. The actual variables are as follows:

- First, *i* is a measure of the solute's ionization. If the solute disassociates, the number of individual particles in the solution increases, which decreases the solute potential. Organic and nonpolar molecules have an *i* value of 1. Ions have an *i* value dependent on their disassociation into a cation (a positive ion) and anion (a negative ion). NaCl disassociates into 2 ions (Na$^+$ and Cl$^-$) and has an *i* value of 2. $CaCl_2$ separates into 3 ions (Ca2$^+$ and 2Cl$^-$) and has an *i* value of 3.

- C (concentration) is the molarity (moles per liter) of the solution.

- T is the absolute temperature of the system in Kelvin ($^\circ C + 273$).

Water passes through the semipermeable cell membrane through aquaporins, while small, uncharged substances, such as oxygen and carbon dioxide, diffuse freely. Water's passage is an example of channel-mediated facilitated diffusion, which is simply diffusing across the cell membrane with the help of a tunnel-like protein. In addition to channel-mediated diffusion, there is also carrier-mediated facilitated diffusion. Carrier proteins, such as the one that allows glucose into the cell, change shape upon solute binding. Facilitated diffusion is a type of passive transport and does not require energy (ATP).

Active Transport

Some transport requires energy for molecules to cross the membrane. Protein pumps are transmembrane proteins that use ATP to change their conformations so that they can force substances against their concentration gradient. For example, the sodium-potassium pump uses ATP to move three sodium (Na+) ions out of the cell and two potassium (K+) ions into the cell. This maintains the steep voltage gradient across the membrane in neurons and muscle cells so that the intracellular environment is more negative than the extracellular one. This membrane potential, along with the concentration gradient, creates an electrochemical gradient.

Another type of active transport, the proton pump, is critical to photosynthesis and respiration through its maintenance of proton gradients across the thylakoid and cristae, respectively.

Endocytosis and exocytosis are examples of active transport required for the bulk movement of substances. They require vesicles, which are plasma membrane-bound delivery sacs. The different types of endocytosis are all similar in that the membrane folds inward so that it pinches off to create a vesicle that carries substances into the cell. The contents then travel to the lysosome, which is an acidic organelle that contains digestive enzymes. The lysosome breaks down the materials so that the cell can use them. Types of endocytosis include:

- Phagocytosis, or "cellular eating," occurs when a cell engulfs large particles and internalizes them by using vacuoles. This only happens in specialized cells such as immune cells.

- Pinocytosis, or "cellular drinking," occurs when a cell engulfs droplets of extracellular fluid and surrounds them with vesicles. This happens routinely in animal cells.

- Receptor-mediated endocytosis occurs when a ligand binds to a receptor protein that initiates a signal transduction cascade.

Endocytosis

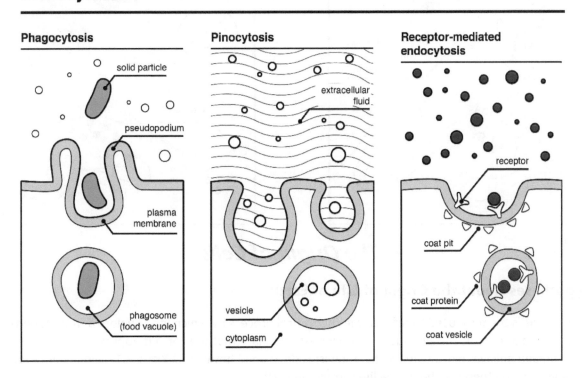

Exocytosis is the excretion of substances via vesicles. Neurotransmitters are excreted via exocytosis at synapses between neurons. Vesicles also release digestive enzymes, hormones, and cellular wastes.

Exocytosis is also used when membrane proteins processed in the Golgi apparatus attach to vesicles that deliver the proteins to their membrane destination.

Exocytosis

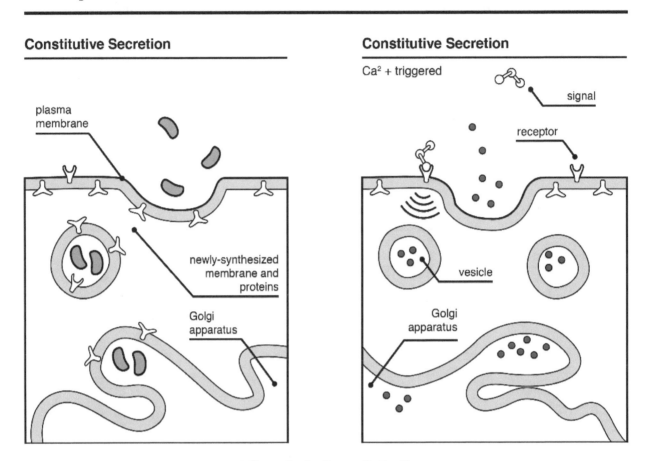

The Origin of Cells

Hypotheses About the Origin of Life of Earth

Abiogenesis is the term used to refer to the theoretical process by which life developed from non-living matter, such as organic compounds. Many scientists estimate that the Earth is approximately 4.5 billion years old and believe that the first living organisms developed between 3.8 and 4.1 billion years ago. There are several theories about how life formed on Earth.

One theory on the origin of life involves the creation of organic compounds from a combination of minerals from the sea and ideal atmospheric conditions. In the deep sea, hydrothermal vents release minerals from the Earth's interior along with hot water. Water from the alkaline vents on the sea floor has a high pH and provides a stable environment for organic compounds. In addition, theorists propose that the Earth's atmosphere had reducing qualities that could produce organic compounds from simpler molecules. They further believe that the warm atmosphere above volcanoes was ideal for this synthesis of organic compounds. The **primordial soup theory** proposes that larger and more complex compounds were made over time from these original organic compounds, eventually forming living organisms.

Panspermia is the concept that life came to Earth from other areas of the universe. It hypothesizes that meteoroids, asteroids, and other small objects from space landed on Earth and transferred microorganisms to the Earth's surface. This theory proposes that there were seeds of life everywhere in the universe, and when these seeds were brought to Earth, the conditions were ideal for living organisms to develop and flourish.

Protocells are small, round groups of lipids hypothesized to be responsible for the origin of life. **Vesicles** are fluid-filled compartments enclosed by a membrane-like structure. They form spontaneously when lipids, such as protocells, are added to water and have important features for the creation of living organisms, such as dividing on their own to form new vesicles and absorbing material around them. When vesicles encounter lipids, the lipids form a bilayer around the vesicle, similar to a plasma membrane of a cell. According to scientists, these cells were able to encapsulate minerals and organic molecules around them. Some of the clay that covered the primordial Earth was believed to be covered in RNA, which the vesicles could encapsulate. These simple behaviors are believed to have given rise to more complex behaviors, such as simple cell metabolism, and began resembling true cells, now known as protocells. As the cells interacted with each other, theorist propose that larger living organisms were created.

Evolutionists propose that life on Earth began with RNA. Although RNA is a genetic material, it can also be an enzyme-like catalyst, known as a **ribozyme**. Ribozymes can catalyze chemical reactions and self-replicate to make complementary copies of short pieces of RNA. According to the theory of evolution, vesicles that carried RNA could then divide and have replicated RNA in its daughter cell, increasing the amount of genetic material in the environment. These daughter cells would have been protocells. Evolutionists believe that the RNA inside of them were likely used as templates to then create more stable DNA strands. It's proposed that then from the formation of DNA, the origin of life and more complex living organisms began. Ultimately, however, it seems evident that the first cells arose from nonliving materials.

In 2015, scientists Sergei Maslov and Alexei Tkachenko expanded on this theory. They believe that the self-replicating model was cyclical and went through different phases during the day and night. During the day, the polymers would float freely, while at night, the polymer chains would join together to form longer polymers using a template, a process called **template-assisted ligation**. They believe that although the chains could join without a template, the use of a template is more efficient and reproducible for preserving the original sequences. According to these two scientists, these phases occurred at different times due to changes in the environment, such as with temperature, pH, and salinity. These factors then regulated whether the polymers would come together or float apart.

Scientific Evidence of Origin Theories

Over the years, many scientific experiments have attempted to prove that different theories about the origin of life are true. In 1953, the Miller-Urey chemical experiment simulated what they believe to be the atmospheric conditions of early Earth. It was believed that the atmosphere contained water, methane, ammonia, and hydrogen. Scientists Stanley Miller and Harold Urey showed that an electrical spark, like a bolt of lightning, helped catalyze the creation of complex organic compounds from simpler ones. They hypothesized that the complex molecules would then react with each other and the simple compounds to form even more compounds, such as formaldehyde, hydrogen cyanide, glycine, and sugars to produce life.

In support of self-replicating RNA, several scientists have tried to create the shortest RNA chain possible that can replicate itself. In the 1960s, Sol Spiegelman created a short RNA chain consisting of 218 bases

that was able to replicate itself with an enzyme from a 4500 base bacterial RNA. In 1997, Manfred Eigen was able to further degrade a large RNA chain to only approximately 50 bases, which was the minimum length needed to bind a replication enzyme. Similarly, researchers at the J. Craig Venter Institute have used engineering techniques to try to create prokaryotic cells with as few genes as possible to figure out the minimal requirements for life. In 1995, they started with a microbe with the smallest genome known to humans with 470 genes and were able to take away one gene at a time to leave only 375 essential genes.

Endosymbiosis and the Origin of Eukaryotic Cells

There are three major domains of life: Eukaryota, Archaea, and Bacteria. **Eukaryota** includes all organisms made up of one or more cells that contain a cell nucleus and organelles enclosed by a membrane. **Archaea** comprises single-celled organisms called prokaryotes, which means that they do not have a cell nucleus or organelles bound by membranes. Bacteria is also made up of prokaryotic cells but unlike the species of Archaea, they do not have genes or metabolic pathways. **Viruses** are microscopic parasites that can only live and reproduce within a host body. They are often even smaller than bacteria. Eukaryotes, archaea, bacteria, and viruses have a symbiotic relationship. The **endosymbiosis theory** of evolution states that eukaryotes developed from larger prokaryotes engulfing smaller prokaryote cells without breaking them up. The small prokaryotes provided the larger prokaryotes with extra energy and eventually developed into mitochondria and chloroplasts.

Endosymbiotic Theory

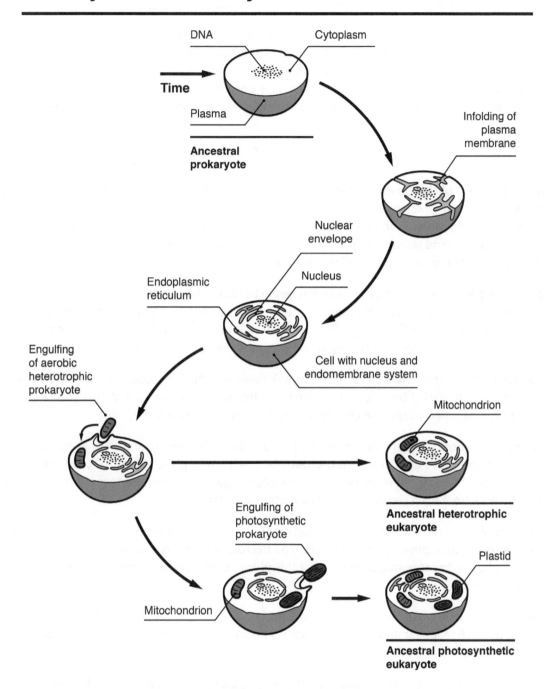

Cell Division

Cellular Reproduction

Cellular reproduction is the process that cells use to divide into two new cells. The ability of a multi-cellular organism to generate new cells to replace dying and damaged cells is vital for sustaining its life. Bacteria reproduce via **binary fission**, which is a simpler process than eukaryotic division because it doesn't involve splitting a nucleus and doesn't have a web of proteins to pull chromosomes apart.

Bacteria copy their DNA in a process called DNA replication, grow, and then the replicated DNA moves to either side, and two new cells are made.

There are two processes by which a eukaryotic cell can divide: mitosis and meiosis. In mitosis, the daughter cells produced from parental cell division are identical to each other and the parent. Meiosis produces genetically unique haploid cells due to two stages of cell division. Meiosis produces haploid cells, or gametes (sperm and egg cells), which only have one set of chromosomes. Humans are diploid because we have two sets of chromosomes—one from each parent. Somatic (body) cells are all diploid and are produced via mitosis.

Mitosis

Mitosis is the division of the genetic material in the nucleus of a cell, and is immediately followed by **cytokinesis**, which is the division of the cytoplasm of the cell. The two processes make up the mitotic phase of the cell cycle. Mitosis can be broken down into five stages: prophase, prometaphase, metaphase, anaphase, and telophase. Mitosis is preceded by interphase, where the cell spends the majority of its life while growing and replicating its DNA.

Prophase: During this phase, the mitotic spindles begin to form. They are made up of centrosomes and microtubules. As the microtubules lengthen, the centrosomes move farther away from each other. The nucleolus disappears and the chromatin fibers begin to coil up and form chromosomes. Two sister **chromatids**, which are two identical copies of one chromosome, are joined together at the centromere.

Prometaphase: The nuclear envelope begins to break down and the microtubules enter the nuclear area. Each pair of chromatin fibers develops a **kinetochore**, which is a specialized protein structure in the middle of the adjoined fibers. The chromosomes are further condensed.

Metaphase: The microtubules are stretched across the cell and the centrosomes are at opposite ends of the cell. The chromosomes align at the **metaphase plate**, which is a plane that is exactly between the two centrosomes. The centromere of each chromosome is attached to the kinetochore microtubules that are stretching from each centrosome to the metaphase plate.

Anaphase: The sister chromatids break apart, forming individual chromosomes. The two daughter chromosomes move to opposite ends of the cell. The microtubules shorten toward opposite ends of the cell as well. The cell elongates and, by the end of this phase, there is a complete set of chromosomes at each end of the cell.

Telophase: Two nuclei form at each end of the cell and nuclear envelopes begin to form around each nucleus. The nucleoli reappear and the chromosomes become less condensed. The microtubules are broken down by the cell and mitosis is complete.

Cytokinesis divides the cytoplasm by pinching off the cytoplasm, forming a cleavage furrow, and the two daughter cells then enter interphase, completing the cycle.

Plant cell mitosis is similar except that it lacks centromeres, and instead has a microtubule organizing center. Cytokinesis occurs with the formation of a cell plate.

The Cell Cycle has Checkpoints

Mitosis has several checkpoints. If a cell exits the cell cycle, it enters a phase called G0 that consists of non-dividing cells like neurons. G1 checkpoints prepare and commit cells to entering the cell cycle. S phase proofreads and corrects mistakes. There is also a G2 checkpoint: as a cell progresses through G1, S, and G2, the cyclin protein accumulates. When it becomes abundant, it binds with a cyclin dependent kinase (cdk) to form **Maturation Promoting Factor (MPF)**, an activating protein complex that facilitates mitosis. As mitosis is completed, the cyclin is degraded, the MPF complex disassembles, and the cell cycle begins once again.

Cell Cycle in Eukaryotes: Interphase

The eukaryotic cell cycle involves four phases: growth (G_1), replication of DNA (S), preparation for cell division (G_2), and cell division (M). The three phases preceding cell division are collectively called **interphase**.

S Phase: DNA replication

A cell must duplicate its DNA prior to cell division so that each daughter cell receives the full set of genetic instructions. To do so, cells use many enzymes. Helicase binds to an origin of replication, separates the two DNA strands, and forms a replication bubble. It then travels along the double-stranded DNA, unzipping it. Topoisomerase is an enzyme downstream of helicase that prevents supercoiling of the DNA. It does this by snipping the DNA, swiveling around it, and then pasting the strands back together. Single-stranded binding proteins hold the DNA strands apart at the replication fork.

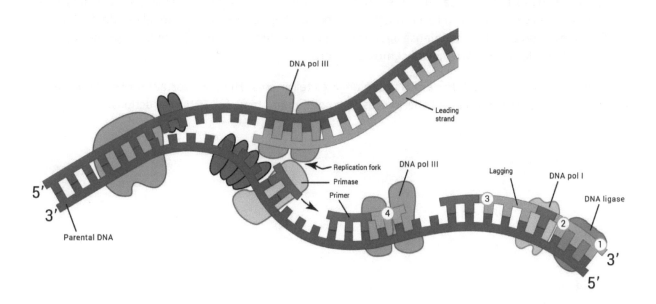

29

The two DNA strands are copied differently.

Leading Strand – Continuous	Lagging Strand – Interrupted
In this strand, DNA is replicated continuously as the helicase travels. DNA polymerase III is the elongating enzyme that copies the DNA in a complementary fashion. Before DNA polymerase III can attach to the DNA and begin, an enzyme called *primase* must insert a small RNA primer for the polymerase to bind to because DNA polymerase III is unable to bind directly to DNA.	This synthesis is similar to the leading strand in that primase deposits a primer for DNA polymerase III binding. However, because DNA cannot be created in the 3' to 5' direction, DNA polymerase III replicates DNA in small 5' to 3' chunks called *Okazaki fragments*. The pieces are eventually connected after DNA polymerase I replaces the RNA primer with DNA. An enzyme called *ligase* glues the small fragments together.

The cell has several levels to correct mistakes during replication:

- DNA polymerase can proofread as it goes.
- Other enzymes can mismatch repair if polymerase fails to recognize mutations.
- Nucleotide excision repair is a higher-tier corrective mechanism. It involves an enzyme called nuclease snipping out the mutation, followed by ligase gluing in the correct sequence.

Telomeres are non-coding, repeating sequences of DNA at the ends of each chromosome. The ends of the chromosome lack a 3' end for DNA polymerase to attach to and replace the RNA primer. As a result, DNA polymerase is unable to bind and, at each division, a little bit of the ends of each chromosome are un-replicated and lost. When the telomeres shorten to nonexistence (at senescence), it's a signal to the cell to end its cell cycling days and go through the process of apoptosis or intentional cell death. This could have evolved as a natural defense against cancer since older cells, which are more prone to mutations, are killed off before they become dangerous.

Germ cells have the enzyme **telomerase** that elongates telomeres. This is important for reproductive purposes so that **gametes** (egg and sperm cells) can continue to have full DNA instructions.

M Phase

Following DNA synthesis is the G2 phase, where the cell assembles the machinery necessary for cell division. Cell division then occurs in several stages, as described before and summarized again below:

PHASE	PHASE EVENTS	ANIMAL CELL DIAGRAM	PLANT CELL DIAGRAM
Prophase	Nucleus disappears and DNA condenses into chromosomes. DNA is already wrapped around histone proteins, and it continues to supercoil until it looks like the letter X. Sister chromatids on either side of the X are identical.		
Metaphase	Chromosomes line up in the cell's center. Kinetochore microtubules extend from animal centrosomes that contain centrioles (organizing centers) on either side of the cell and attach to the centromeres (repeating sequences in the middle of the chromosomes). Nonkinetochore proteins elongate animal cells. This massive protein orchestra is collectively called the spindle apparatus. Plants lack centrioles but have microtubule organizing centers.		
Anaphase	Kinetochore microtubules shorten/are pulled in and sister chromatids are separated and move to each daughter cell.		
Telophase and Cytokinesis	Nuclei reform and chromosomes decondense within them. **Animals:** actin and myosin microfilaments pinch off the cytoplasm at the cleavage furrow to form two new cells **Plants:** cell plate (new cell wall) forms between daughter cells and extends to divide into two new cells		

Mitosis occurs for many reasons:

- Development and growth of an organism
- Differentiation and specialization in multicellular organisms
- Replacement of cells with damage or rapid turnover

Cancer cells are the result of inappropriate cell division and occur when cells are unresponsive to checkpoint regulation and growth factor signals. Normal cells are density dependent and anchorage dependent, but cancer cells lose these properties.

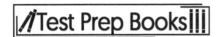
Meiosis

Meiosis is a type of cell division in which the parent cell has twice as many sets of chromosomes as the daughter cells into which it divides. Although the first stage of meiosis involves the duplication of chromosomes, similar to that of mitosis, the parent cell in meiosis divides into four cells, as opposed to the two produced in mitosis.

Meiosis has the same phases as mitosis, except that they occur twice: once in meiosis I and again in meiosis II. The diploid parent has two sets of homologous chromosomes, one set from each parent. During meiosis I, each chromosome set goes through a process called **crossing over**, which jumbles up the genes on each chromatid. In anaphase one, the separated chromosomes are no longer identical and, once the chromosomes pull apart, each daughter cell is haploid (one set of chromosomes with two non-identical sister chromatids). Next, during meiosis II, the two intermediate daughter cells divide again, separating the chromatids, producing a total of four total haploid cells that each contains one set of chromosomes.

Practice Questions

1. The image below shows an experiment that was conducted with the disaccharide maltose. The initial solutions on the left were measured after 30 minutes using a graduated cylinder and a disaccharide indicator. The dotted line represents a membrane that is only permeable to water. Which of the following best explains the results?

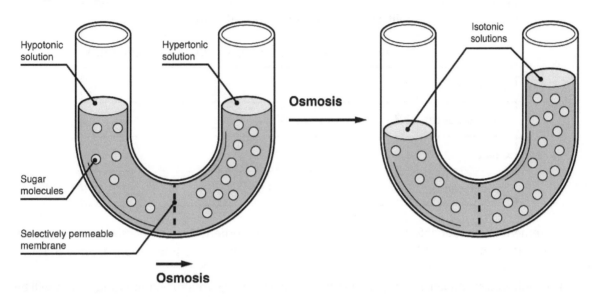

a. The initial solution on the left of the u-tube is hypotonic, so the solvent will move to the right until both sides are isotonic.
b. The initial solution on the right of the u-tube has a greater water potential than the solution on the left, so the solvent will move to the right.
c. The solute moves from the left side of the u-tube to the right until equilibrium is reached.
d. The initial solution on the left of the u-tube has a smaller water potential than the solution on the left, so the solute will move in.

2. What is the solute potential of an open 0.15 M calcium chloride solution at 23 °C?
   a. -3.7
   b. -11.1
   c. -0.29
   d. -0.86

3. Microvilli in the small intestine serve an important function by increasing nutrient absorption. The three conditions in the table all affect blood sugar circulation in different ways, which are all indirectly related to absorption. The more efficient the absorption, the more sugar enters the bloodstream. Type I diabetes involves little to no insulin production, which causes very high blood sugar if not treated with exogenous insulin. People with celiac disease have trouble digesting gluten. Hypoglycemia results in very low blood sugar.

Microvilli Size in Different Groups of Individuals		
	Average Length (μm)	Average Width and Height (μm)
Unaffected	4.7	0.8
Type I diabetes	3.3	1.7
Celiac	1.1	1.7
Hypoglycemia	5.0	0.5

Which statement is a reasonable conclusion given the data recorded? The sample size was very small, and although these numbers do not necessarily reflect that of the whole population, assume that they do.

a. The individuals with celiac disease have the smallest surface area to volume ratio, which will be beneficial because it will make nutrient absorption more efficient.

b. The individuals with hypoglycemia have a larger surface area to volume ratio, which will be beneficial because it will make nutrient absorption more efficient.

c. The unaffected individuals have the largest surface area to volume ratio, which will be beneficial because it will make nutrient absorption more efficient.

d. The unaffected individuals have the smallest surface area to volume ratio, which will be beneficial because it will make nutrient absorption more efficient.

4. What organelle is the site of protein synthesis?

a. Nucleus

b. Smooth ER

c. Ribosome

d. Lysosome

5. Why would an aquaporin allow the passage of glycerol ion but not a hydronium ion?

a. Glycerol is closer in size to water than a hydronium ion.

b. A hydronium ion is slightly larger than water and is prevented from passing through the aquaporin.

c. The passage of glycerol is allowed because it has a positive charge on one of the oxygen atoms.

d. A hydronium ion is not allowed to pass because the extra proton results in a charged molecule.

The structures of glycerol, water, and a hydronium ion are shown in the figure below.

Glycerol

OH

HO⎯⎯⎯⎯⎯OH

Water

H⎯O⎯H

Hydronium Ion

+

H⎯O⎯H

H

Answer Explanations

1. A: Hypotonic means that a solution is less concentrated than the one that it is being compared to and therefore, has a lower concentration of solute and a higher concentration of water. The solution on the left is hypotonic, and due to the principles of passive transport, water will move from high concentration to low concentration and will therefore, move from the left side to the right side until both sides are isotonic, making Choice A correct. Solute concentrations are unchanged between the two sides after equilibrium is reached, illustrating that the membrane is semipermeable and that the disaccharide maltose is too large to pass through, throwing out Choices C and D that refer to solute movement. Water potential decreases as solute increases, and because the solution on the right side of the u-tube has more solute, it therefore has a lower water potential. This makes Choice B incorrect.

2. B: Solute potential is determined by the following equation.

$$\Psi_s = -iCRT$$

The variables represent the following:

- First, i is the ionization constant and accounts for the fact that ionic compounds dissociate. Calcium chloride ($CaCl_2$) will dissociate into Ca^{+2} and $2Cl^{-1}$, meaning that for every calcium chloride molecule, there will be 3 ions circulating. Remember that solute potential is the pressure needed (provided by solutes) that will prevent the inward flow of water. Solute potential decreases as solute increases. Similar to when obstacles are added to an obstacle course, it makes movement more difficult. In this particular equation, i equals 3.

- C represents the molar concentration, in this case 0.15 $CaCl_2$. The higher the concentration, the lower the solute potential.

- T represents temperature and it needs to be in Kelvin (°C + 273). In this equation, it will be equal to 23 + 273, which equals 296.

- R is the pressure constant, which is .0831 $L \times bar \div (mole \times K)$.

Substitute to find the solution

$$\Psi_s = -3(0.15 \; mole \; L) \times .0831 \; L \; bar \div mole \times K \times (296K) = -11.1 \; bar$$

The other answer choices are calculations using either the wrong ionization constant or lacking the Kelvin conversion.

3. B: The question is based on absorption efficiency and surface area to volume ratio. The larger the ratio, the better the cell will be at absorption because there is more surface area for absorption to occur. The first step is to calculate the surface area to volume ratios of the different groups. This can be done by likening the microvilli structure to a rectangular prism or cylinder, which will not give exact answers, but will serve as a model that will reflect the trend. The following calculations use a rectangular prism model for calculations.

Surface area of a rectangular prism minus the side where it is attached to the small intestine is:

$length \times width \times 4 + the \; outer \; face$, which is $width \times height$

The volume of the rectangular prism is:

$$length \times width \times height$$

To find the surface area to volume ratio, divide the surface area by the volume. The highest surface area to volume ratio is the most efficient microvilli, meaning it is the one absorbing the most nutrients.

Group	Surface Area	Volume	Ratio
	L x W x 4 + W x H	L x W x H	
Unaffected	15.7	3.0	5:1
Type I diabetes	25.3	9.5	3:1
Celiac	10.4	3.2	3:1
Hypoglycemia	10.3	1.3	8:1

Choices *C* and *D* are wrong because the unaffected group does not have the largest or smallest surface area to volume ratio. Choice *A* is wrong because while the celiac group does have the smallest surface area to volume ratio, that makes it the least efficient, not the most. Choice *B* is the best choice because a large surface area to volume ratio results in more efficient transport.

4. C: Proteins are synthesized on ribosomes. The ribosome uses messenger RNA as a template and transfer RNA brings amino acids to the ribosome where they are synthesized into peptide strands using the genetic code provided by the messenger RNA.

5. D: Choice *A* is not correct since glycerol contains fourteen atoms and water contains three. A hydronium ion is closer in size to the water molecule. However, the size of a hydronium ion must not be a factor since glycerol, the larger molecule, is allowed passage through the aquaporin. Therefore, Choice *B* is not a likely choice. Choice *C* indicates that glycerol contains a charge on one of the oxygen atoms. If a molecule is charged, typically, the word "ion," "cation," or "anion" would be specified. The "-ol" ending of the word "glycerol" indicates that the compound is an alcohol. Therefore, Choice *C* is not correct. The structure of a hydronium ion suggests that the charge is the reason why the ion cannot pass through the aquaporin. A hydronium ion is the only molecule that specifically contains a positive charge. Choice *D* is the correct answer choice. Furthermore, the charge is the reason for the hydronium ion's exclusion, which indicates that aquaporin contains amino groups along the passage wall that specifically participate in weak bonding (for example, dipole-dipole) but prevent charged molecules from passing.

Molecular Biology

Molecules to Metabolism

Carbon is the Foundation of Biological Molecules

Carbon is the foundation of organic molecules because it has the ability to form four covalent bonds and long polymers. The four organic compounds are lipids, carbohydrates, proteins, and nucleic acids.

- **Lipids** are critical for cell membrane structure, long-term energy storage, and to help form some steroid hormones such as testosterone and cholesterol.

- **Carbohydrates** are important as a medium for energy storage and conversion, but also have structural importance. Cellulose (a monomer of glucose) provides structure for plant cell walls. Chitin provides structure for fungi and animals with exoskeletons (such as crabs and lobsters), and peptidoglycan is a carbohydrate/protein hybrid that forms the cell walls of some prokaryotes.

- **Proteins** are important because enzymes regulate all chemical reactions, but there are also many cell membrane proteins important for structure, transport, and communication.

- **Nucleic acids** include DNA, the genetic instructions of organisms, and RNA, which is the molecule responsible for turning those instructions into products.

All of these organic compounds require the elements carbon (C), hydrogen (H), and oxygen (O), which enter the food chain through the glucose that it produces through photosynthesis. Some compounds contain phosphorus (P), sulfur (S), and nitrogen (N) as well. Elements such as phosphorus and nitrogen diffuse into the roots of plants from the external environment, are incorporated into organic compounds, and are distributed to other organisms through symbiotic relationships or food webs.

Bioenergetics

Bioenergetics refers to the flow of energy within a biological system and is primarily focused on how chemical energy in the macronutrients from food (i.e., carbohydrates, proteins, fats) is converted into biologically-usable forms of energy (i.e., substrates) that organisms can use to perform work.

Catabolism
Catabolism is the process of breaking large molecules into smaller molecules to make energy available to the organism. For example, carbohydrates are catabolized to provide fuel for exercise and daily living. Catabolism also can involve the breakdown of muscle tissue during periods of heavy training volumes, low caloric intake, or high stress.

Anabolism
Anabolism is the process of restructuring or building larger compounds from catabolized materials, such as assembling amino acids into structural proteins, which are needed to maintain homeostasis and to generate new muscle tissue.

Metabolism is the sum total of the chemical processes that occur within a cell for the maintenance of life. It includes both the synthesizing and breaking down of substances. A **metabolic pathway** begins

with a molecule and ends with a specific product after going through a series of reactions, often involving an enzyme at each step. **Catabolic pathways** are metabolic pathways in which energy is released, as complex molecules are broken down into simpler molecules. In contrast, **anabolic pathways** use energy to build complex molecules out of simpler molecules. With cell metabolism, it is important to remember the first law of thermodynamics: Energy can be transformed, but it cannot be created or destroyed. Therefore, the energy released in a cell by catabolic pathways is used up in anabolic pathways.

The reactions that occur within metabolic pathways are classified as either exergonic reactions or endergonic reactions. **Exergonic reactions** end in a release of free energy, while endergonic reactions absorb free energy from the surroundings. **Free energy** is the portion of energy in a system, such as a living cell, that can be used to perform work, such as a chemical reaction. It is denoted as the capital letter G, and the change in free energy from a reaction or set of reactions is denoted as delta G (ΔG). When reactions do not require an input of energy, they are said to occur spontaneously. Exergonic reactions are considered **spontaneous** because they result in a negative delta G ($-\Delta$G), where the products of the reaction have less free energy within them than the reactants. **Endergonic reactions** require an input of energy and result in a positive delta G ($+\Delta$G), with the products of the reaction containing more free energy than the individual reactants. When a system no longer has free energy to do work, it has reached **equilibrium**. Since cells always need to do work, they are no longer alive if they reach true equilibrium.

Cells balance their energy resources by using the energy from exergonic reactions to drive endergonic reactions forward, a process called **energy coupling**. **Adenosine triphosphate**, or ATP, is a molecule that is an immediate source of energy for cellular work. When it is broken down, it releases energy used in endergonic reactions and anabolic pathways. ATP breaks down into **adenosine diphosphate**, or ADP, and a separate phosphate group, releasing energy in an exergonic reaction. As ATP is used up by reactions, it is also regenerated by having a new phosphate group added onto the ADP products within the cell in an endergonic reaction.

Enzymes are special proteins that help speed up metabolic reactions and pathways. They do not change the overall free energy release or consumption of reactions; they just make the reactions occur more quickly by lowering the required activation energy. Enzymes are designed to act only on specific substrates. Their physical shape fits snugly onto their matched substrates, so enzymes only speed up reactions that contain the substrates to which they are matched.

Adenosine Triphosphate (ATP)
ATP is a high-energy molecule used for muscle contractions, endergonic reactions, movement, and other life-sustaining metabolic processes. ATP is an **intermediate molecule** (consisting of three primary parts—an adenine, a ribose, and three phosphates in a chain) that allows energy to transfer from exergonic to endergonic and catabolic to anabolic reactions. ATP is generated and replenished in skeletal muscles by three energy systems: phosphagen, glycolytic, and oxidative.

ATP Hydrolysis
Hydrolysis is a general term for any chemical reaction that breaks a chemical body via the addition of water. ATP hydrolysis splits the ATP molecule into adenosine diphosphate (ADP) and usable energy. The enzyme **adenosine triphosphatase (ATPase)** is the catalyst for the hydrolysis of ATP.

The following equation shows the reactants (left of arrow), enzyme (middle), and products (right of arrow) for ATP hydrolysis:

$$ATP + H_2O \longleftarrow ATPase \longrightarrow ADP + P_i + H^+ + Energy$$

When ATP undergoes hydrolysis, ADP (containing two phosphate groups), an inorganic phosphate molecule, a hydrogen ion, and free energy are produced.

ATPase is the enzyme responsible for catalyzing the breakdown of ATP to ADP. The dephosphorylation reaction results in the release of energy, which is then used to carry out other chemical reactions.

Myosin ATPase catalyzes ATP hydrolysis, providing the energy for cross-bridge recycling.

Calcium ATPase is the enzyme that provides the energy used to regulate calcium movement, by pumping it into the sarcoplasmic reticulum.

The ATP - ADP Cycle

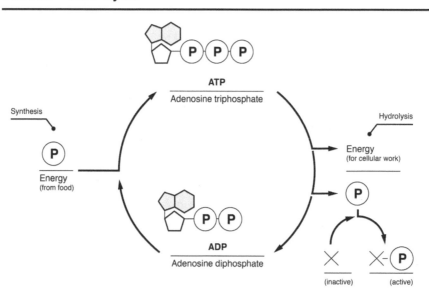

Sodium-Potassium ATPase controls the sodium potassium concentration gradient in the sarcolemma after depolarization to maintain the cellular resting potential. For every two K^+ ions pumped in, there are three NA^+ ions pumped out of the cell.

Adenosine Monophosphate (AMP)
Adenosine Monophosphate (AMP) results from ADP hydrolysis, which cleaves the second phosphate group, leaving one.

Water

While water is not an organic compound (meaning it does not contain carbon), it is a critical molecule for life. While it is **covalent**, in that the oxygen shares electrons with two hydrogen atoms, it also has

polar bonds, meaning that it is slightly charged and borderline ionic. Oxygen has an atomic number of 8, and hydrogen has an atomic number of 1. Think of the oxygen nucleus as having eight positive magnets in it, and the hydrogen nucleus of having one. Now think of hydrogen having one lone electron circling around it. That electron is going to be sucked toward the more electronegative, or "powerful," oxygen nucleus, making the oxygen slightly negative. The hydrogen proton will be exposed on one side, and that part of it will be attracted to any negative oxygen neighbors. This intermolecular attraction between the hydrogen proton and a partially negative neighbor is called a **hydrogen bond**. Hydrogen bonds are *not* ionic; they are intermolecular attractions. The image of liquid water here shows the hydrogen bonds as dotted lines. These hydrogen bonds give water many unique properties, including the following:

- **Cohesion** is when water sticks to itself because of the attraction between molecules, as observed in dew.

- **Adhesion** is when water sticks to other structures, as observed in graduated cylinders when sides of the meniscus stick to the walls.

- **High surface tension** is when water can support some solids. Solids are supposed to be less dense than liquid, but if there is a large enough body of water, there will be a significant "sticky" film at the top due to the cohesive forces between water molecules. Heavy things can penetrate the film and sink, but lighter objects such as leaves and ice, even though they are solid, will not be able to overcome the attractive forces between the water molecules and will float.

- **High boiling point** is a result of the great amount of kinetic energy required to overcome water's cohesive forces and vaporize.

- **High freezing point** is a result of water's attractive forces, enabling it to be arranged much more easily into the specific hexagonal arrangements that form ice, permitting it to solidify at a higher temperature.

- **Solid ice** has a very low density due to the space between the lattice/crystalline structure, which explains why ice is less dense than water, as seen in the image below.

Hydrogen bonds

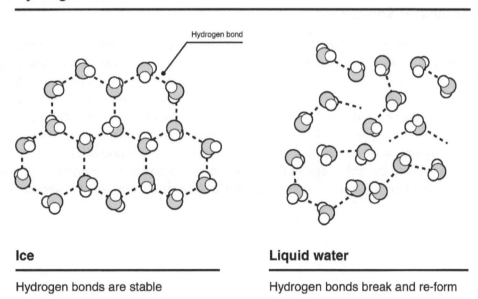

Ice
Hydrogen bonds are stable

Liquid water
Hydrogen bonds break and re-form

These unique properties of water sustain life. Ice floating prevents bodies of water from freezing entirely solid because it insulates the liquid water underneath. This insulation keeps the water below it fluid and allows organisms to survive. Adhesion and cohesion of water in the xylem of plants allows water to travel great distances against the force of gravity.

Cohesion

Cohesion is the interaction of many of the same molecules. In water, cohesion occurs when there is hydrogen bonding between water molecules. Water molecules use this bonding ability to attach to each other and can work against gravity to transport dissolved nutrients to the top of a plant. A network of water-conducting cells can push water from the roots of a plant up to the leaves.

The cohesive behavior of water also causes surface tension. If a glass of water is slightly overfull, water can still stand above the rim. This is because of the unique bonding of water molecules at the surface—they bond to each other and to the molecules below them, making it seem like it is covered with an impenetrable film. A raft spider can actually walk across a small body of water due to this surface tension.

Adhesion

Adhesion is the linking of two different substances. Water molecules can form a weak hydrogen bond with, or adhere to, plant cell walls to help fight gravity.

Water Has High Specific Heat Capacity

Another important property of water is its ability to moderate temperature. Water can moderate the temperature of air by absorbing or releasing stored heat into the air. Water has the distinctive capability of being able to absorb or release large quantities of stored heat while undergoing only a small change in temperature. This is because of the relatively high specific heat of water, where specific heat is the amount of heat it takes for one gram of a material to change its temperature by 1 degree Celsius. The specific heat of water is one calorie per gram per degree Celsius, meaning that for each gram of water, it takes one calorie of heat to raise or lower the temperature of water by 1 degree Celsius.

Water is a Universal Solvent

The polarity of water molecules makes it a versatile solvent. Ionic compounds, such as salt, are made up of positively and negatively charged atoms, called cations and anions, respectively. Cations and anions are easily dissolved in water because of their individual attractions to the slight positive charge of the hydrogen atoms or the slight negative charge of the oxygen atoms in water molecules. Water molecules separate the individually charged atoms and shield them from each other so they don't bond to each other again, creating a homogenous solution of the cations and anions. Nonionic compounds, such as sugar, have polar regions, so are easily dissolved in water. For these compounds, the water molecules form hydrogen bonds with the polar regions (hydroxyl groups) to create a homogenous solution. Any substance that is attracted to water is termed hydrophilic. Substances that repel water are termed **hydrophobic**.

Carbohydrates and Lipids

Carbohydrates

Carbohydrates are molecules consisting of carbon, oxygen, and hydrogen atoms and serve many biological functions, such as storing energy and providing structural support. The monomer of a carbohydrate is a sugar monosaccharide. Some biologically important sugars include glucose, galactose, and fructose. These monosaccharides are often found as six-membered rings and can join together by a process known as condensation to form a disaccharide, which is a polymer of two sugar units.

Polysaccharides are long chains of monosaccharides. Their main biological functions are to store energy and provide structure. Starch and cellulose are both polymers of glucose made by plants, but their functions are very different because the structure of the polymers are different. Plants make starch to store the energy that is produced from photosynthesis, while cellulose is an important structural component of the cell wall. These two molecules differ in how the molecules of glucose are bonded together. In starch, the bond between the two glucose molecules is a high-energy alpha bond that is easily hydrolyzed, or broken apart. The bonds between the glucose molecules in cellulose are beta bonds. Being stiff, rigid, and hard to hydrolyze, these bonds give the cell wall structural support. Other important carbohydrates include glycogen, which provides energy storage in animal cells, and chitin, which provides structure to arthropod exoskeletons.

Lipids

Lipids are a class of biological molecules that are **hydrophobic**, meaning they don't mix well with water. They are mostly made up of large chains of carbon and hydrogen atoms, termed **hydrocarbon chains**. When lipids mix with water, the water molecules bond to each other and exclude the lipids because they are unable to form bonds with the long hydrocarbon chains. The three most important types of lipids are fats, phospholipids, and steroids.

Fats are made up of two types of smaller molecules: glycerol and fatty acids. **Glycerol** is a chain of three carbon atoms, with a hydroxyl group attached to each carbon atom. A **hydroxyl group** is made up of an oxygen and hydrogen atom bonded together. **Fatty acids** are long hydrocarbon chains that have a backbone of sixteen or eighteen carbon atoms. The carbon atom on one end of the fatty acid is part of a carboxyl group. A **carboxyl group** is a carbon atom that uses two of its four bonds to bond to one oxygen atom (double bond) and uses another one of its bonds to link to a hydroxyl group. Fats are made by joining three fatty acid molecules and one glycerol molecule.

Glycerol   **Fatty Acid**

Phospholipids are made of two fatty acid molecules linked to one glycerol molecule. A phosphate group is attached to a third hydroxyl group of the glycerol molecule. A phosphate group has an overall negative charge and consists of a phosphate atom connected to four oxygen atoms.

Phospholipids have an interesting structure because their fatty acid tails are hydrophobic, but their phosphate group heads are hydrophilic. When phospholipids mix with water, they create double-layered structures, called **bilayers**, that shield their hydrophobic regions from water molecules. Cell membranes are made of phospholipid bilayers, which allow the cells to mix with aqueous solutions outside and inside, while forming a protective barrier and a semi-permeable membrane around the cell.

Steroids are lipids that consist of four fused carbon rings. The different chemical groups that attach to these rings are what make up the many types of steroids. Cholesterol is a common type of steroid found

in animal cell membranes. Steroids are mixed in between the phospholipid bilayer and help maintain the structure of the membrane and aids in cell signaling.

Proteins

Proteins are essential for most all functions in living beings. The name protein is derived from the Greek word *proteios*, meaning *first* or *primary*. Proteins are molecules that consist of carbon, hydrogen, oxygen, nitrogen, and other atoms, and they have a wide array of functions. The monomers that make up proteins are amino acids. All amino acids have the same basic structure. They contain an amine group (-NH), a carboxylic acid group (-COOH), and an R group. The **R group**, also called the **functional group**, is different in each amino acid.

The **functional groups** give the different amino acids their unique chemical properties. There are twenty naturally occurring amino acids that can be divided into groups based on their chemical properties. Glycine, alanine, valine, leucine, isoleucine, methionine, phenylalanine, tryptophan, and proline have nonpolar, hydrophobic functional groups. Serine, threonine, cysteine, tyrosine, asparagine, and glutamine have polar functional groups. Arginine, lysine, and histidine have charged functional groups that are basic, and aspartic acid and glutamic acid have charged functional groups that are acidic.

A **peptide bond** can form between the carboxylic-acid group of one amino acid and the amine group of another amino acid, joining the two amino acids. A long chain of amino acids is called a polypeptide or protein. Because there are so many different amino acids and because they can be arranged in an infinite number of combinations, proteins can have very complex structures. There are four levels of protein structure. Primary structure is the linear sequence of the amino acids; it determines the overall structure of the protein and how the functional groups are positioned in relation to each other, as well as how they interact. Secondary structure is the interaction between different atoms in the backbone chain of the protein. The two main types of secondary structure are the alpha helix and the beta sheet. Alpha helices are formed when the N-H of one amino-acid hydrogen bonds with the C=O of an amino acid four amino acids earlier in the chain.

The functional groups of certain amino acids—including methionine, alanine, uncharged leucine, glutamate, and lysine—make the formation of alpha helices more likely. The functional groups of other amino acids, such as proline and glycine, make the formation of alpha helices less likely. Alpha helices are right-handed and have 3.6 residues per turn. Proteins with alpha helices can span the cell membrane and are often involved in DNA binding. Beta sheets are formed when a protein strand is stretched out, allowing for hydrogen bonding with a neighboring strand.

Similar to alpha helices, certain amino acids have an increased propensity to form beta sheets. Tertiary protein structure forms from the interactions between the different functional groups and gives the protein its overall geometric shape. Interactions that are important for tertiary structure include hydrophobic interactions between nonpolar side groups, hydrogen bonding, salt bridges, and disulfide bonds. Quaternary structure is the interaction that occurs between two different polypeptide chains and involves the same properties as tertiary structure. Only proteins that have more than one chain have quaternary structure.

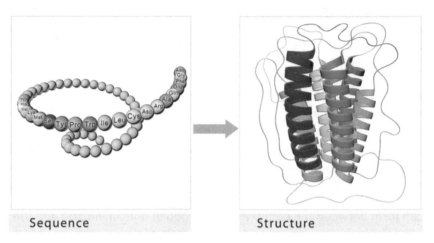

Sequence Structure

Enzymes

Enzymes are a class of catalysts instrumental in biochemical reactions, and in most, if not all, instances, they are proteins. Like all catalysts, enzymes increase the rate of a chemical reaction by providing an alternate path that requires less activation energy. Enzymes catalyze thousands of chemical reactions in the human body. Enzymes possess an active site, which is the part of the molecule that binds the reacting molecule, or **substrate.** The "lock and key" analogy is used to describe the substrate key fitting precisely into the active site of the enzyme lock to form an enzyme-substrate complex.

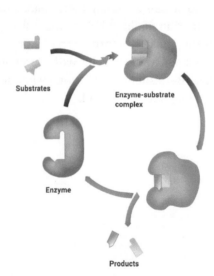

Substrates

Enzyme-substrate complex

Enzyme

Products

Many enzymes work in tandem with cofactors or coenzymes to catalyze chemical reactions. **Cofactors** can be either inorganic (not containing carbon) or organic (containing carbon). Organic cofactors can be

either coenzymes or prosthetic groups tightly bound to an enzyme. **Coenzymes** transport chemical groups from one enzyme to another. Within a cell, coenzymes are continuously regenerating, and their concentrations are held at a steady state.

Several factors—including temperature, pH, and concentrations of the enzyme and substrate—can affect the catalytic activity of an enzyme. For humans, the optimal temperature for peak enzyme activity is approximately body temperature at 98.6 ^{0}F, while the optimal pH for peak enzyme activity is approximately 7–8. Increasing the concentrations of either the enzyme or substrate will also increase the rate of reaction, up to a certain point.

The activity of enzymes can be regulated. One common type of enzyme regulation is termed **feedback inhibition**, which involves the product of the pathway inhibiting the catalytic activity of the enzyme involved in its manufacture.

DNA and RNA

Nucleic Acids

Nucleic acids have two important duties in the body. As monomers, they are crucial for energy transfer. As polymers, they are a fundamental component of genetic material. Nucleotides are the monomer units that assemble to form nucleic acids. Nucleotides have three components: a nitrogenous base and a phosphate functional group, both of which are attached to a five-carbon (pentose) sugar. There are two classes of nitrogenous bases, purines and pyrimidines. The two types of purines are guanine (G) and adenine (A), while the three types of pyrimidines are thymine (T), cytosine (C), and uracil (U). The two types of pentose sugars are deoxyribose and ribose. Nucleotides containing deoxyribose are termed deoxyribonucleic acid (DNA). DNA utilizes guanine, adenine, cytosine, and thymine as its nitrogenous bases. Nucleotides containing ribose are termed ribonucleic acid (RNA). RNA utilizes guanine, adenine, cytosine, and uracil as its nitrogenous bases.

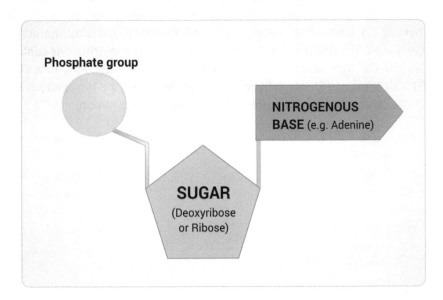

RNA

Ribonucleic acid (RNA) plays crucial roles in protein synthesis and gene regulation. RNA is made of nucleotides consisting of ribose (a sugar), a phosphate group, and one of four possible nitrogen bases—adenine (A), cytosine (C), guanine (G), and uracil (U). RNA utilizes the nitrogenous base uracil in place of the base thymine found in DNA. Another difference between RNA and DNA is that RNA is typically found as a single-stranded structure, while DNA typically exists in a double-stranded structure.

RNA can be categorized into three major groups—messenger RNA (mRNA), ribosomal RNA (rRNA), and transfer RNA (tRNA). Messenger RNA (mRNA) transports instructions from DNA in the nucleus of a cell to the areas responsible for protein synthesis in the cytoplasm of a cell. This process is known as **transcription**. Transfer RNA (tRNA) deciphers the amino acid sequence for the construction of proteins found in mRNA. Both tRNA and ribosomal RNA (rRNA) are found in the ribosomes of cells. Ribosomes are responsible for protein synthesis. The process is known as **translation**, and both tRNA and rRNA play crucial roles. Both translation and transcription are further described below.

DNA

Deoxyribonucleic acid (DNA) contains the genetic material that is passed from parent to offspring. It contains specific instructions for the development and function of a unique eukaryotic organism. The great majority of cells in a eukaryotic organism contains the same DNA.

The majority of DNA can be found in the cell's nucleus and is referred to as **nuclear DNA**. A small amount of DNA can be located in the mitochondria and is referred to as **mitochondrial DNA**. Mitochondria provide the energy for a properly functioning cell. All offspring inherit mitochondrial DNA from their mother. James Watson, an American geneticist, and Frances Crick, a British molecular biologist, first outlined the structure of DNA in 1953.

The structure of DNA visually approximates a twisting ladder and is described as a double helix. DNA is made of nucleotides consisting of deoxyribose (a sugar), a phosphate group, and one of four possible nitrogen bases—thymine (T), adenine (A), cytosine (C), and guanine (G). It is estimated that human DNA contains three billion bases. The sequence of these bases dictates the instructions contained in the DNA making each species singular. The bases in DNA pair in a particular manner—thymine (T) with adenine (A) and guanine (G) with cytosine (C). Weak hydrogen bonds between the nitrogenous bases ensure easy uncoiling of DNA's double helical structure in preparation for replication.

Structure and Function of DNA

Deoxyribonucleic acid (DNA) is life's instruction manual. It is double stranded and directional, meaning it can only be transcribed and replicated from the *3' end* to the *5' end*.

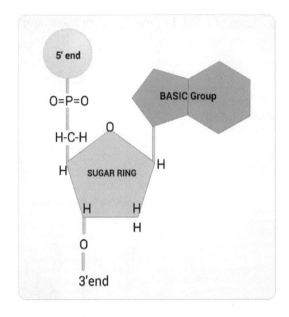

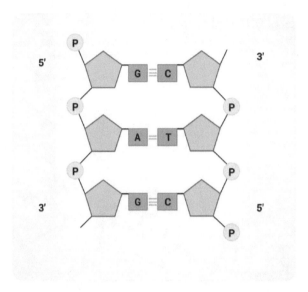

DNA as a Monomer	DNA as a Polymer
A nucleotide is composed of a five-carbon sugar with a Phosphate Group off of the 5th carbon and a Nitrogen Base off of the 1st carbon. DNA and RNA are different because DNA contains deoxyribose sugar while RNA contains ribose sugar. Also, the nitrogen base thymine in DNA is replaced by uracil in RNA.	The two strands are antiparallel, meaning they are read in opposite directions. The bases guanine and cytosine are complementary and held together by three hydrogen bonds, and the bases adenine and thymine are complementary and held together by two hydrogen bonds. Weak hydrogen bonding between bases allows DNA to be opened easily for transcription and replication.

Nitrogen bases come in two varieties:

- *One-Ringed Pyrimidines*: Cytosine and Thymine (DNA)/Uracil (RNA)
- *Two-Ringed Purines*: Adenine and Guanine

Purines and Pyrimidines

The five bases in DNA and RNA can be categorized as either pyrimidine or purine according to their structure. The **pyrimidine** bases include cytosine, thymine, and uracil. They are six-sided and have a single ring shape. The **purine** bases are adenine and guanine, which consist of two attached rings. One ring has five sides and the other has six. When combined with a sugar, any of the five bases become nucleosides. Nucleosides formed from purine bases end in "osine" and those formed from pyrimidine bases end in "idine." Adenosine and thymidine are examples of nucleosides. Bases are the most basic components, followed by nucleosides, nucleotides, and then DNA or RNA.

DNA Replication, Transcription, and Translation

DNA Replication

Replication refers to the process during which DNA makes copies of itself. Enzymes govern the major steps of DNA replication.

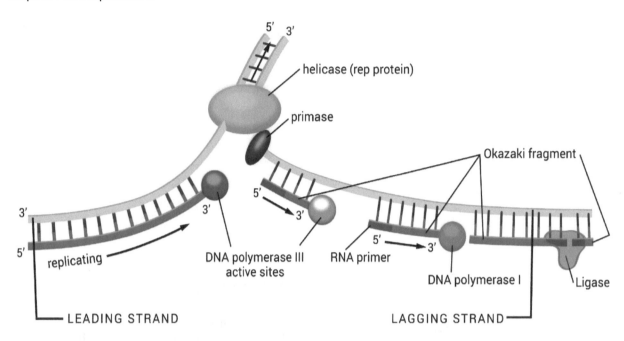

The process begins with the uncoiling of the double helix of DNA. **Helicase**, an enzyme, accomplishes this task by breaking the weak hydrogen bonds uniting base pairs. The uncoiling of DNA gives rise to the replication fork, which has a Y-shape. Each separated strand of DNA will act as a template for the production of a new molecule of DNA. The strand of DNA oriented toward the replication fork is called the **leading strand** and the strand oriented away from the replication fork is named the **lagging strand**.

Replication of the leading strand is continuous. **DNA polymerase**, an enzyme, binds to the leading strand and adds complementary bases. Replication of the lagging strand of DNA, on the other hand, is discontinuous. DNA polymerase produces discontinuous segments, called **Okazaki fragments**, which are later joined together by another enzyme, **DNA ligase**. To start the DNA synthesis on the lagging strand, the enzyme **primase** lays down a strip of RNA, called an **RNA primer**, to which the DNA polymerase can bind. As a result, two clones of the original DNA emerge from this process. DNA replication is considered **semiconservative** due to the fact that half of the new molecule is old, and the other half is new.

Transcription

Transcription refers to a portion of DNA being copied into RNA, specifically mRNA. It represents the first crucial step in gene expression.

Prokaryotic transcription consists of the following three stages:

- **Initiation**: RNA polymerase binds to the start site on the RNA template strand, which is upstream in the promotor region of the gene.

- **Elongation**: Synthesis of the new complementary RNA strand begins by the action of RNA polymerase as it works its way down the gene, growing the new strand.

- **Termination**: The new RNA chain is completed and the RNA polymerase is released at the stop point, which is downstream on the gene in the terminator region.

Eukaryotic transcription is more complicated:

- **Eukaryotic Initiation**: Eukaryotes contain advanced promoters, which are DNA sequences upstream of genes that recruit transcription factors, proteins that facilitate or block RNA polymerase binding. The TATA box is where RNA polymerase II binds. RNA polymerase then untwists the double helix of DNA by breaking weak hydrogen bonds between its nucleotides.

- **Eukaryotic Elongation**: Once DNA is untwisted, RNA polymerase travels down the strand reading the DNA sequence and adding complementary nitrogenous bases. With the assistance of RNA polymerase, the pentose sugar and phosphate functional group are added to the nitrogenous base to form a nucleotide. This is similar to prokaryotic elongation in that it generates a 5' to 3' complementary RNA transcript.

- **Eukaryotic Termination**: A polyadenylation signal (DNA sequence AAUAAA) causes RNA polymerase to release several bases downstream. The RNA transcript is then processed. Upstream of the promoter, past a 5' untranslated region, a "cap" is added (like a telomere, but only a few repeated Guanines). Downstream of the polyadenylation signal, a poly(A) tail is added. In addition to changes at the end of each transcript, areas within the transcript are further processed through splicing. Spliceosomes (large complexes of RNA and proteins) remove intermittent noncoding regions called introns, and the exons (coding sequences) are joined together. The introns are spliced out so that only the exons are part of the final transcript. Alternative splicing (when one transcript is spliced in different ways and creates different proteins) makes this process even more complicated. Lastly, the weak hydrogen bonds uniting the DNA-RNA complex are broken to free the newly formed mRNA. The mRNA travels from the nucleus of the cell out to the cytoplasm of the cell where translation occurs.

Codons are groups of three nucleotides on the messenger RNA, and can be visualized as three rungs of a ladder. A codon has the code for a single amino acid. There are 64 codons but 20 amino acids. More than one combination, or triplet, can be used to synthesize the necessary amino acids. For example, AAA (adenine-adenine-adenine) or AAG (adenine-adenine-guanine) can serve as codons for lysine. These groups of three occur in strings, and might be thought of as frames. For example, AAAUCUUCGU, if read in groups of three from the beginning, would be AAA, UCU, UCG, which are codons for lysine, serine, and serine, respectively. If the same sequence was read in groups of three starting from the second position, the groups would be AAU (asparagine), CUU (proline), and so on. The resulting amino acids would be completely different. For this reason, there are start and stop codons that indicate the beginning and ending of a sequence (or frame). AUG (methionine) is the start codon. UAA, UGA, and UAG, also known as ocher, opal, and amber, respectively, are stop codons.

Translation

Translation is the process of generating protein from RNA. Translation in prokaryotes is far simpler than in eukaryotes. They only have one circular chromosome (so far fewer genes), and they don't have DNA within a nucleus. They also don't process transcripts. Translation can even occur simultaneously with transcription on the same piece of RNA. In fact, many different ribosomes can be working on the same transcript at the same time, thus creating a structure called a **polyribosome**.

The ribosomes of bacteria and eukaryotes are similar. A ribosome has two subunits, both made of ribosomal RNA (*rRNA*, which is made by RNA polymerase I in eukaryotes). In eukaryotes, the small subunit and the large subunit assemble in conjunction with the AUG start codon in the mRNA message (DNA is read in three letter "words" called codons).

Before translation can start, a transfer RNA molecule (*tRNA*, which is made by RNA polymerase III) must join the ribosomal complex. tRNAs contain anticodons that are complementary to specific codons on the RNA transcript. On one side, their highly specific anticodon binds to corresponding codons in the mRNA. On the other side, tRNA molecules carry specific amino acids, the monomer of proteins. Once a tRNA molecule has released its amino acid, an enzyme amino-acyl-tRNA synthetase will join a free-floating tRNA and its corresponding amino acid. As a result, the tRNA can continue to deliver fresh amino acids to the ribosomes.

The point of translation is that, every time a tRNA molecule drops off an amino acid, it contributes to an emerging protein. Only when a stop codon is reached will the ribosome disassemble, thus releasing the assembled protein.

		Second Base in Codon		
	U	**C**	**A**	**G**
U	UUU ⎫ Phe UUC ⎭ UUA ⎫ Leu UUG ⎭	UCU ⎫ UCC ⎬ Ser UCA UCG ⎭	UAU ⎫ Tyr UAC ⎭ UAA Stop UAG Stop	UGU ⎫ Cys UGC ⎭ UGA Stop UGG Trp	U C A G
C	CUU ⎫ CUC ⎬ Leu CUA CUG ⎭	CCU ⎫ CCC ⎬ Pro CCA CCG ⎭	CAU ⎫ His CAC ⎭ CAA ⎫ Gln CAG ⎭	CGU ⎫ CGC ⎬ Arg CGA CGG ⎭	U C A G
A	AUU ⎫ AUC ⎬ Ile AUA AUG Met or Start	ACU ⎫ ACC ⎬ Thr ACA ACG ⎭	AAU ⎫ Asn AAC ⎭ AAA ⎫ Lys AAG ⎭	AGU ⎫ Ser AGC ⎭ AGA ⎫ Arg AGG ⎭	U C A G
G	GUU ⎫ GUC ⎬ Val GUA GUG ⎭	GCU ⎫ GCC ⎬ Ala GCA GCG ⎭	GAU ⎫ Asp GAC ⎭ GAA ⎫ Glu GAG ⎭	GGU ⎫ GGC ⎬ Gly GGA GGG ⎭	U C A G

First Base in Codon / Third Base in Codon

Note in the chart above that each codon codes for a specific amino acid, even specifically coding for stop codons. There are many codon combinations, but only 20 amino acids. The redundancy in codon/amino acid pairs is due to the third base—or "**wobble**" position—of tRNA. The first two bases bind so strongly that sometimes the third base does not play much of an active role in connecting the codon and anticodon.

The ribosome has three tRNA landing spots. The first tRNA binds to the start codon in the "P" site, but every tRNA that follows lands at the "A" site. The growing amino acid chain is added to the amino acid in the A site (connecting via a peptide bond). The "E" site is where the naked tRNA exits after it has removed its amino acid chain.

Once a stop codon is reached, the amino acid chain leaves through an exit tunnel. At this point, it is an immature protein that is not properly folded and might be sent to the Golgi for modification.

Transcription Regulation

Transcriptional regulation in prokaryotes occurs via operons. An **operon** is a DNA sequence comprised of related genes which are clustered behind a single promoter that regulates them. Within the promoter is an operator, which is like an on/off switch.

As discussed earlier, eukaryotic transcription initiation involves transcription factors. Different transcription factors are present in different cell types, and certain sequences of DNA enhance protein binding. This results in an altered DNA spatial arrangement and exposure to RNA polymerase. Some transcription factors facilitate transcription while some block it. The presence or absence of these proteins in certain cells is one way that gene production is regulated.

Another way transcription is regulated in eukaryotes is by gene accessibility. DNA is wrapped around histone proteins in complexes called **nucleosomes**. These nucleosomes coil and supercoil, which make portions of the genome inaccessible. Modification of histone tails in nucleosomes open and close regions of DNA, making them either more or less available for the protein binding of transcription factors and RNA polymerase. When the chromatin is in its closed, coiled conformation, called **heterochromatin**, transcription is repressed because the genes are inaccessible. In contrast, in the **euchromatin** formation, the chromatin is open and uncoiled, which allows RNA polymerase to access the genes, which activates transcription.

Essentially, open chromatin (*euchromatin*) has acetylated histone tails with few methyl groups. **Heterochromatin** is the opposite, with histone tails being highly methylated and deacetylated so that DNA is closed and condensed.

MicroRNAs (*miRNAs*) and small interfering RNAs (*siRNAs*) can also regulate gene production. This is done by degrading certain transcripts or blocking their translation, and sometimes even altering chromatin structure.

It should be noted that there is far more DNA functionality than gene expression. DNA also contains promoters and enhancers to regulate gene expression, centromeres, telomeres, transposable elements, and other sequences.

Transposable Elements
These are literally "jumping genes" that relocate within an organism's genome. Prokaryotes and eukaryotes contain **transposons**. This supports the idea that they have a significant effect on

biodiversity and arose through common ancestry among all living organisms. DNA sequences prone to transposon trespassing are predictable, repeated bases that can extend from 300 to 30,000 nucleotides.

Within an organism, transposons relocate and might leave the targeted segment behind. A transposon can make a copy of the DNA or simply cut out the sequence. Either way, transposase is a critical enzyme involved in the process.

Retrotransposons are elements that copy the targeted segment into an RNA transcript. The enzyme retrotransposase is then used to copy the intermediate back into DNA before inserting it, thus copying part of the genome and moving it elsewhere.

It's believed that all human ancestors once had brown eyes until a single individual had a mutation which coded for blue eyes. Once that allele entered the gene pool, it was passed to generations of offspring. Today it's fairly common to see a blue-eyed individual. Thus, changes in alleles can change phenotypes.

Variations are Introduced by the Imperfect Nature of DNA Replication and Repair

Despite replication's proofreading, mutations do happen. Mutations can be either point mutations or chromosomal mutations, and both increase genetic variation.

Point mutations involve a change in DNA sequence and are due to an addition, deletion, inversion, or substitution error.

Insertion and deletion mutations cause frame-shift mutations. This means that every following codon will be read incorrectly, massively changing the primary structure of the protein. Due to the redundancy of codon/amino acid pairing resulting from the wobble position, some substitution mutations cause no change in the protein sequence. The result is something called a **silent mutation**. All point mutations (including substitution and inversion) can cause gene malfunction.

Chromosomal Mutations

Sometimes there are **chromosomal mutations** in DNA replication and mitosis. The following types of mutations can occur:

- **Deletion:** A section of a chromosome is removed.
- **Duplication:** A section of a chromosome is repeated.
- **Inversion:** A chromosome is rearranged within itself.
- **Translocation:** Chromosome pieces mix or combine with other chromosomes.

Not only can mutations lead to changes in an individual's phenotype, but they can also contribute to reproductive isolation and speciation which have effects on a much larger scale.

In metaphase I in meiosis, homologous chromosomes align in the center of the cell. The law of independent assortment states that it is random whether a mother's or father's chromosomes are on the left or right. Since humans have 23 chromosomes, there are 2^{23} (over 8 million) possible variations. Independent assortment, random fertilization, and recombination contribute greatly to genetic diversity.

Cell Respiration

Cellular Respiration in Eukaryotes

All organisms, whether autotrophic or heterotrophic, use food to produce ATP in a process called respiration. **Cellular respiration** is the metabolic process that converts energy from nutrients into ATP and waste products. This will be explained in detail later. Prokaryotes use proteins on their cell membrane to perform respiration, while eukaryotes have specialized structures called mitochondria to do it.

Chloroplasts and mitochondria are the organelles responsible for all energy conversion in eukaryotic cells. All eukaryotes have mitochondria, but only plants and green algae have chloroplasts as well.

Glycolysis

The first step of breaking down glucose to make energy is called glycolysis (literally "glucose-splitter"), and it occurs in the cytosol of cells.

$$C_6H_{12}O_6 + 2ATP \rightarrow 2C_3H_4O_3 + 4ATP + 2NADH$$

$$Glucose + activation\ energy \rightarrow 2\ Pyruvate + energy + 2\ electron\ carriers$$

As shown in the previous formula, the overall goal of glycolysis is to break glucose in half and into 2 pyruvate molecules. In doing so, it peels off high-energy electrons that were contained in glucose. Two pairs of electrons (stored in phosphate groups) and two hydrogen atoms are invested into the electron carrier NAD+ that behaves just like the electron carrier in photosynthesis by shuttling electrons from one process to the next.

Glycolysis requires a 2ATP energy investment to proceed to completion, and it produces 4ATP via substrate level phosphorylation. This net gain of 2ATP is a small percentage of the total energy produced in aerobic respiration.

In the absence of oxygen, the 2ATP produced in glycolysis is the only energy gain there is, and fermentation, or anaerobic respiration, will initiate to recycle the electron carrier NAD$^+$. The two chief types of anaerobic respiration/fermentation are lactic acid fermentation and alcohol fermentation. When muscle cells have exceeded their aerobic capacity, they go into anaerobic respiration, which produces lactic acid and 2 net ATP. Yeast undergoes alcohol fermentation, producing carbon dioxide, ethyl alcohol, and 2 net ATP.

Glycolysis is performed via the following steps:

- Glucose is converted into Glucose-6-Phosphate (G6P) via the enzyme hexokinase (note that any enzyme ending in -kinase indicates that a phosphate group is going to be donated). Energy is lost because ATP loses a phosphate group and is converted into ADP in the process of creating G6P.

- G6P is then rearranged, turning from a hexagonal figure to a pentagonal figure, and is configured into Fructose-6-Phosphate (F6P) by an enzyme called phosphoglucose isomerase.

- F6P is converted into Fructose-1,6-Bisphosphate (F1,6BP) by phosphofructokinase, donating a phosphate group from ATP and turning it into ADP, losing energy.

- F1,6BP is then broken down into two 3-carbon molecules by the enzyme aldolase. These molecules are DHAP (dihydroxyacetone phosphate) and G3P (glucose-3-phosphate). DHAP acts as a kind of brake on glycolysis. If there is too much energy in the body, this reaction favors DHAP. However, if there is not enough energy in the body, this reaction favors G3P to go on to continue glycolysis.

- G3P is then converted into 1,3-Bisphosphoglycerate (1,3-BPG) via glyceraldehyde phosphate dehydrogenase. During this conversion, NADP+ is converted into NADPH.

- 1,3-BPG becomes 3-Phosphoglycerate (3-PG) via the enzyme phosphoglycerate kinase). Energy here is gained in the form of ATP, in which ADP gains a phosphate group through this enzyme.

- 3-PG is then mutated by phosphoglyceromutase into 2-Phosphoglycerate (2-PG).

- 2-PG is converted into phosphoenolpyruvate (PEP) via the enzyme enolase. Water is created via this process.

- Finally, PEP is converted into pyruvate through the enzyme pyruvate kinase, in which energy is again gained by the donation of a phosphate group to ADP to create ATP. Pyruvate is then entered into the Citric Acid (Krebs) cycle.

The Citric Acid Cycle

If oxygen is present in a eukaryotic organism, the remainder of the process of respiration will occur inside the mitochondria via the Krebs cycle and oxidative phosphorylation. The main goal of the Citric Acid cycle is to take pyruvate and break it down, producing NADH and $FADH_2$.

The citric acid cycle has eight steps. Remember that glycolysis produces two pyruvate molecules from each glucose molecule. Each pyruvate molecule oxidizes into a single acetyl-CoA molecule, which then enters the citric acid cycle. Therefore, two citric acid cycles can be completed and twice the number of ATP molecules are generated per glucose molecule.

Step 1: Acetyl-CoA adds a two-carbon acetyl group to an oxaloacetate molecule and produces one citrate molecule.

Step 2: Citrate is converted to its isomer, isocitrate, by removing one water molecule and adding a new water molecule in a different configuration.

Step 3: Isocitrate is oxidized and converted to α-ketoglutarate. A carbon dioxide (CO_2) molecule is released and one NAD+ molecule is converted to NADH.

Step 4: α-Ketoglutarate is converted to succinyl-CoA. Another carbon dioxide molecule is released and another NAD+ molecule is converted to NADH.

Step 5: Succinyl-CoA becomes succinate by the addition of a phosphate group to the cycle. The oxygen molecule of the phosphate group attaches to the succinyl-CoA molecule and the CoA group is released. The rest of the phosphate group transfers to a guanosine diphosphate (GDP) molecule, converting it to

guanosine triphosphate (GTP). GTP acts similarly to ATP and can actually be used to generate an ATP molecule at this step.

Step 6: Succinate is converted to fumarate by losing two hydrogen atoms. The hydrogen atoms join a flavin adenine dinucleotide (FAD) molecule, converting it to $FADH_2$, which is a hydroquinone form.

Step 7: A water molecule is added to the cycle and converts fumarate to malate.

Step 8: Malate is oxidized and converted to oxaloacetate. One lost hydrogen atom is added to an NAD molecule to create NADH. The oxaloacetate generated here then enters back into step one of the cycle.

At the end of glycolysis and the citric acid cycles, four ATP molecules have been generated. The NADH and $FADH_2$ molecules are used as energy to drive the next step of oxidative phosphorylation.

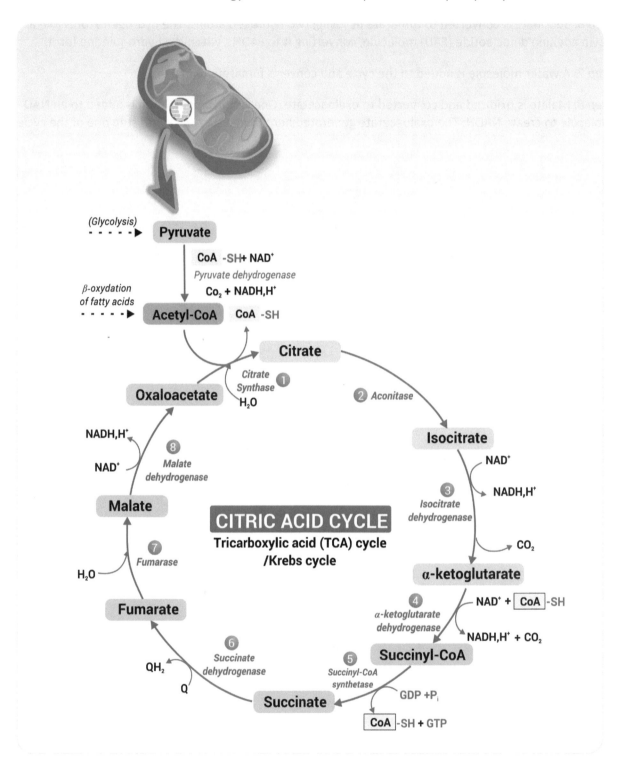

Including the intermediate stage, where pyruvate enters, two spins of the cycle (one for each pyruvate) produces the following:

- 2 CO_2 (from intermediate)
- 2 NADH (from intermediate)
- 4 CO_2
- 6 NADH
- 2 $FADH_2$
- 2 ATP (or GTP) via substrate-level phosphorylation. GTP is guanosine tri-phosphate, which is analogous to ATP.

NADH and $FADH_2$ are electron carriers, meaning they donate electrons to the electron transport chain. NADH donates two high-energy electrons and one proton (H+), while $FADH_2$ donates two electrons and two protons (2 H^+). These high-energy electron and proton carriers then go through the process of oxidative phosphorylation, where many ATP molecules are made. Using the energy supplied by the electron transport chain, protons are transported through the integral protein complexes I, III, and IV along the inner mitochondrial membrane. The pumping of these protons out of the mitochondrial matrix across the cristae to the inner membrane space, establishes a concentration gradient. Just like in photosynthesis, this gradient provides the proton motive force to generate ATP when the hydrogen ion later passes through ATP synthase, causing it to spin and convert ADP to ATP.

Oxidative Phosphorylation

Oxidative phosphorylation includes two steps: the electron transport chain and chemiosmosis. The inner mitochondrial membrane has four protein complexes, sequenced I to IV, used to transport protons and electrons through the inner mitochondrial matrix. Two electrons and a proton (H+) are passed from each NADH and $FADH_2$ to these channel proteins, pumping the hydrogen ions to the inner-membrane space using energy from the high-energy electrons to create a concentration gradient. NADH and $FADH_2$ also drop their high-energy electrons to the electron transport chain. NAD+ and FAD molecules in the mitochondrial matrix return to the Krebs cycle to pick up materials for the next delivery. From here, two processes happen simultaneously:

- **Electron Transport Chain:** In addition to complexes I to IV, there are two mobile electron carriers present in the inner mitochondrial membrane, called ubiquinone and cytochrome C. At the end of this transport chain, electrons are accepted by an O_2 molecule in the matrix, and water is formed with the addition of two hydrogen atoms from chemiosmosis.

- **Chemiosmosis:** This occurs in an ATP synthase complex that sits next to the four electron transporting complexes. ATP synthase uses facilitated diffusion (passive transport) to deliver protons across the concentration gradient from the inner mitochondrial membrane to the matrix. As the protons travel, the ATP synthase protein physically spins, and the kinetic energy generated is invested into phosphorylation of ADP molecules to generate ATP. Oxidative phosphorylation produces twenty-six to twenty-eight ATP molecules, bringing the total number of ATP generated through glycolysis and cellular respiration to thirty to thirty-two molecules.

Energy Extraction from Cellular Respiration

Aerobic respiration uses oxygen and produces 30–32 ATP molecules using the mitochondria. Only a few of the ATP are generated via substrate-level phosphorylation in glycolysis and the Krebs cycle; the vast majority of ATP is generated through the electron transport chain and chemiosmosis.

The exact number of ATP molecules made per glucose molecule varies. Glycolysis and the Krebs cycle each produce a net gain of 2 ATP/GTP via substrate-level phosphorylation. Oxidative phosphorylation is more difficult to calculate. Each electron carrier NADH produces around 2.5 ATP, while $FADH_2$ produces around 1.5 ATP. These are not whole numbers because there is not a direct relationship between electron transport and phosphorylation — they are two different processes. One has to do with electrons traveling down the chain to the final electron acceptor: oxygen. The other has to do with the movement of hydrogen ions. Finally, some of the work done by oxidative phosphorylation might be distributed to other cellular processes because respiration does not exist in a vacuum.

Another example of the flexibility of energy production by respiration is seen in thermoregulation. It was discussed earlier that endothermic organisms have ways to regulate body heat, including shivering and sweating. Another is by using an uncoupling protein in the cristae during hibernation. A mitochondrial protein called uncoupling protein 1 (UCP1) in brown fat cells hijacks the proton motive force by preventing them from entering the ATP synthase and creating ATP. Instead, UCP1 moves the protons by increasing the permeability of the inner mitochondrial membrane, using energy from the proton gradient to be dissipated as heat. This helps keep hibernating animals warm without creating unneeded ATP, which helps animals conserve energy and keep their metabolic rates low.

In addition to two pyruvate molecules produced by glycolysis, six molecules of NADH and two molecules of flavin adenine dinucleotide ($FADH_2$) are produced and used by the ETC. Hydrogen atoms, transported by NADH and $FADH_2$ to the ETC, are used to produce ATP from ADP. The hydrogen atoms form a proton concentration gradient down the ETC that produces energy required to produce ATP. NADH and $FADH_2$ molecules rephosphorylate ADP to ATP via the ETC with each NADH producing three ATP molecules and $FADH_2$ producing two ATP molecules.

Anaerobic Respiration

Some organisms do not live in oxygen-rich environments and must find alternate methods of respiration. **Anaerobic respiration** occurs in certain prokaryotic organisms. They utilize an electron transport chain similar to the aerobic respiration pathway; however, the terminal acceptor molecule is an electronegative substance that is not O_2. Some bacteria, for example, use the sulfate ion (SO_4^{2-}) as the final electron accepting molecule and the resulting byproduct is hydrogen sulfide (H_2S), instead of water.

Muscle cells that reach anaerobic threshold go through lactic acid respiration, while yeasts go through alcohol fermentation. Both processes only make two ATP.

Photosynthesis

Photosynthesis is the process of converting light energy into chemical energy, which is then stored in sugar and other organic molecules. It can be divided into two stages called the **light reactions** and the **Calvin cycle**. The photosynthetic process takes place in the chloroplast in plants. Inside the chloroplast, there are membranous sacs called **thylakoids**. **Chlorophyll** is a green pigment that lives in the thylakoid membranes, absorbs photons from light, and starts an electron transport chain in order to produce energy in the form of ATP and NADPH. The ATP and NADPH produced from the light reactions are used as energy to form organic molecules in the Calvin cycle.

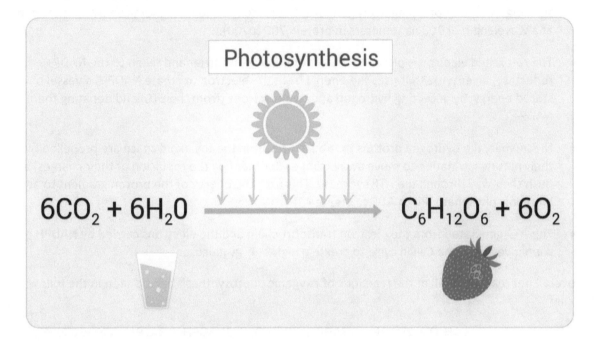

Photosynthesis

$$6CO_2 + 6H_2O \longrightarrow C_6H_{12}O_6 + 6O_2$$

Stage One of Photosynthesis: The Light-Dependent Reaction

Photosynthesis is the complex process that chlorophyll-containing organisms perform to make their own food, which will then be used to create energy through cellular respiration. Chlorophyll is a green pigment responsible for the absorption of a photon (a unit of light), which provides the energy required to begin photosynthesis. Chlorophyll is present in the eukaryotic cells of plants and plant-like protists such as green algae.

Photosynthesis begins with Stage One, also known as the light dependent reaction, which results in the creation of energy in the form of ATP and NADPH. It is light dependent because it requires the energy provided by a photon. It occurs in a complicated series of steps, which are outlined here:

- Light from the sun (in the form of a photon) strikes a molecule of chlorophyll that is embedded within and around the photosystems lodged within the thylakoid membrane of the chloroplast.

- The photon excites an electron located in Photosystem II (PSII), which is the first of four protein complexes within the membrane. PSII absorbs photons with a wavelength of 680 nanometers in protein 680 (p680). This excited electron jumps into the primary electron acceptor in the center of PSII, where it is picked up by an electron carrier. The electron serves to transfer energy.

- Meanwhile, PSII takes a molecule of water and splits it into hydrogen and oxygen, stealing an electron from hydrogen to replace the one it just lost, and releasing oxygen and protons (H+). This process is called hydrolysis.

- The excited electron travels to the next protein complex, the cytochrome complex—the intermediary between PSII and Photosystem I (PSI)—which uses the energy from the electron to pump a proton across the thylakoid membrane into the thylakoid space, creating a positive concentration gradient.

- The electron, having exhausted all of its energy as it moved along the electron transport chain, splitting water and pumping hydrogen ions, travels to PSI and is re-energized by another photon at a wavelength of 700 nanometers in protein 700 (p700).

- The re-excited electron is picked up by another electron carrier and taken to the NADP+ reductase, an enzyme that uses the energy from the electron to create NADPH, a vessel of stored energy, by accepting hydrogen and two electrons (from the ETC) and donating them to NADP+.

- Meanwhile, the hydrogen protons that build up within the thylakoid space are propelled by their natural inclination to move away from each other (via the repulsion of their charges) and push their way through the ATP synthase. This uses the energy of the proton gradient to add an inorganic phosphate (Pi) to ADP to create ATP.

- The ATP generated from the electron transport chain and the electrons carried by NADPH are soon invested in the Calvin cycle to create a high-energy glucose.

The overall net reaction of all of the reactions of oxygenic photosynthesis can be seen in the following formula:

$$2H_2O + 2NADP+ + 3ADP + 3P_i \rightarrow O_2 + 2NADPH + 3ATP$$

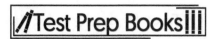

Stage Two of Photosynthesis (The Calvin Cycle): Light-Independent Reactions

The Calvin Cycle

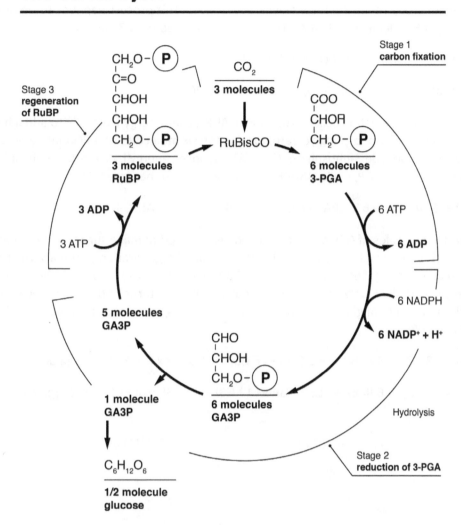

The Calvin cycle is the part of photosynthesis that actually creates glucose. It uses the byproducts ATP and NADPH and the solar energy harnessed in Stage One in a series of events described as follows.

- Carbon dioxide from the atmosphere enters the plant through stomata on the bottoms of its leaves. It then diffuses into the stroma of the chloroplast, located outside the thylakoid membrane.

- In the exergonic reaction called carbon fixation, the CO2 then combines with RuBP, a 5-carbon molecule with two phosphate groups, catalyzed by the enzyme called RuBisCO, the most abundant protein in the world. The addition of a sixth carbon causes the molecule to become unstable, so each molecule of RuBP immediately splits into two 3-carbon molecules with a phosphate group called 3-phosphoglycerate (3-PGA). This process happens three times, resulting in 6 molecules of 3-GPA. (This step is also called the C3 pathway).

1 Turn of the Calvin Cycle: CO_2 + RuBP → 3-PGA

3 Turns of the Calvin Cycle: $3CO_2$ + 3 RuBP → 6 3-PGA

- In an endergonic reaction called reduction, NADPH uses energy from ATP to add a hydrogen to each molecule of 3-phosphoglycerate, turning the six molecules of 3-PGA into a 3-carbon sugar called glyceraldehyde 3-phosphate (G3P). ATP supplies energy by donating a phosphate group (Pi) and becoming ADP, while NADPH loses a hydrogen to become $NADP^+$.

6 ATP + 6 NADPH + 6 3-GPA → 6 G3P + 6ADP + $6P_i$ + $6NADP_+$ + $6H_+$

- Of the six molecules of G3P created, only one is reserved to make sugar, while the other five molecules are reused in an endergonic reaction called regeneration to replace the three used RuBP molecules. It takes two molecules of G3P to make glucose, and since three turns of the Calvin cycle produce only one G3P, this means it takes six turns of the Calvin cycle to make one 6-carbon molecule of glucose. In the following formulas, remember that most of the G3P is used to renew RuBP.

3 Turns = $3 CO_2$ + 3 RuBP → 6 3-PGA → 6 G3P → 1 G3P exit → ½ $C_6H_{12}O_6$ (glucose)

6 Turns = $6 CO_2$ + 6 RuBP → 12 3-PGA → 12 G3P → 2 G3P exit → 1 $C_6H_{12}O_6$ (glucose)

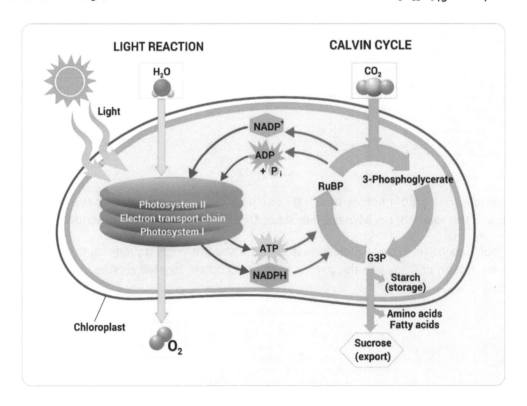

Photorespiration: C3, C4, and CAM Plants

In certain conditions, plants will halt the production of glucose in a process called photorespiration. C3 plants, the most common, are the most efficient in cool, moist climates. In hot or dry conditions, stomata, the miniscule holes in leaves that enable the transfer of liquid and gases between the plant and its environment, close to conserve water. This can be problematic because it reduces the influx of carbon dioxide. Carbon dioxide becomes scarce, and the oxygen byproduct of the hydrolysis during the light-dependent reaction in PSII builds up, unable to escape. When temperatures increase, RuBisCO has a higher affinity for oxygen and that, combined with the higher O_2 to CO_2 ratio caused by the closed stomata, causes it to bind to O_2 instead of CO_2. This means that carbon cannot be fixed to become glucose, but still it uses ATP to burn up energy, essentially undoing the work of the Calvin cycle. Therefore, photorespiration is resource- and energy-draining, and scientists are not entirely sure of its evolutionary significance. Two alternative systems exist in some plants to avoid photorespiration.

C4 plants and CAM plants are found in tropical and desert climates. They have mechanisms to avoid photorespiration that they will employ if resources are low. They both take an alternative route to the Calvin cycle, so they actually have the same sugar-producing endgame as C_3 plants. These pathways that circumvent photorespiration either require extra energy (as with C_4 plants) or are not as efficient (as in CAM plants) as functioning C3 plants, but the fact that they still produce sugar for life-sustaining energy makes them preferable to photorespiration.

C4 (4-carbon) plants evade photorespiration by using a much more efficient enzyme called PEP carboxylase, which has an affinity for carbon dioxide only. RuBisCO is the preferred enzyme for carbon fixation, since PEP carboxylase requires energy. However, if RuBisCO is being blocked by oxygen in low carbon dioxide situations, PEP carboxylase will bind any circulating carbon dioxide and incorporate it into a C4 product. This 4-carbon product (called oxaloacetate, or OAA for short) releases carbon dioxide to the Calvin cycle to be used by RuBisCO in carbon fixation. ATP is invested to convert it to a 3-carbon sugar that can bind to PEP carboxylase and repeat the cycle. Basically, PEP carboxylase is acting as a carbon dioxide pump to keep levels high, enabling it to make the high energy glucose.

CAM plants, typically found in deserts, conserve water by keeping their stomata closed during the hot part of the day; this prevents dehydration through transpiration. The stomata open at night to capture and store carbon dioxide in organic compounds. Like C4 plants, the carbon is fixed into carbon intermediates. During daylight, these compounds are broken up so that released carbon dioxide can bind to RuBisCO and stimulate sugar production via the Calvin cycle. This pathway is beneficial for plants such as cacti that need to conserve water. It is not as efficient due to uncoupling carbon fixation with the rest of the cycle, but for extremely hot environments, it is adaptive.

Practice Questions

1. In aerobic respiration, a hydrogen ion gradient is essential for the production of ATP through oxidative phosphorylation. Which cytoplasmic embedded mitochondria shown below correctly demonstrates the relative concentration of hydrogen ions?

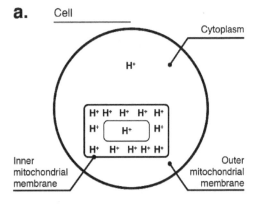

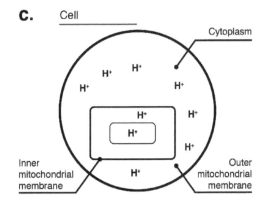

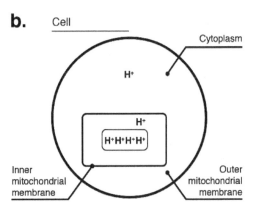

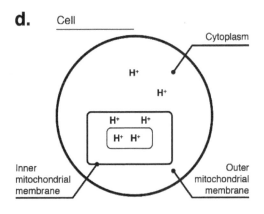

2. Both NADP+ and NAD+ are important for cellular energy conversion. They distribute high-energy electrons to electron transport chains that facilitate the pumping of protons across membranes and couple the action with redox reactions. Which of the following is another similarity between the two molecules?

    a. They both generate ATP by traveling through ATP synthase.

    b. They both carry one proton and a pair of electrons when they are reduced.

    c. They both deliver electrons from either glycolysis or the Krebs cycle to the cristae.

    d. They both are oxidized by electrons energized by the photosystems.

3. What is the term used for the set of metabolic reactions that convert chemical bonds to energy in the form of ATP?
    a. Photosynthesis
    b. Reproduction
    c. Active transport
    d. Cellular respiration

4. What is the role of an allosteric activator?
    a. To bind to the active site of an enzyme and block the binding of a substrate.
    b. To bind to the active site of an enzyme and allow the binding of a substrate.
    c. To bind to an unrelated site of an enzyme to block the binding of a substrate.
    d. To bind to an unrelated site of an enzyme to allow the binding of a substrate.

5. Which of the following is true about an endergonic reaction?

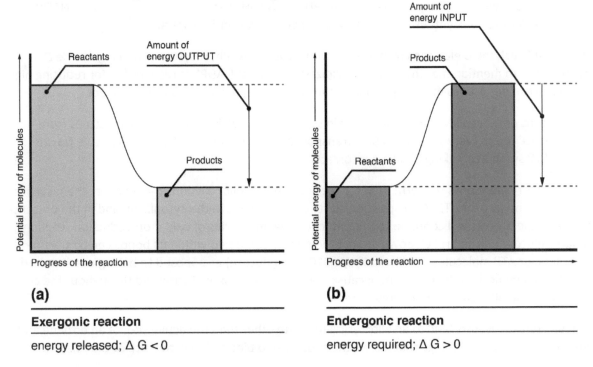

(a)

Exergonic reaction

energy released; $\Delta G < 0$

(b)

Endergonic reaction

energy required; $\Delta G > 0$

    a. The reaction releases energy and decreases entropy.
    b. The reaction absorbs energy and increases entropy.
    c. The reaction releases energy and increases entropy.
    d. The reaction absorbs energy and decreases entropy.

Answer Explanations

1. A: In the process of oxidative phosphorylation, it is critical that there is a concentration gradient where hydrogen ions are at higher levels in the inner membrane space and are at lower levels in the matrix. The mitochondria use energy from the high-energy electrons that NADH and $FADH_2$ are carrying. The energy is used to pump protons across the cristae from the matrix to the inner membrane space. This is significant because it maintains the gradient that is responsible for the facilitated diffusion of the hydrogen ion across the cristae through ATP synthase. ATP synthase spins when the proton goes back through, providing the kinetic energy for the endergonic reaction that creates ATP.

$$ADP + P = ATP$$

2. B: $NADP^+$ and NAD^+ are very similar molecules in that they carry a proton and steal a pair of high-energy electrons when they are reduced to NADPH and NADH. The major difference is that $NADP^+$ is the electron carrier in photosynthesis and NAD^+ is the electron carrier in respiration.

$NADP^+$, not NAD^+, takes electrons from the electrons excited by Photosystem I, making Choice *D* incorrect, not to mention that they are not oxidized. NAD^+, not $NADP^+$, is responsible for reducing high-energy glucose derivatives in glycolysis and in the Krebs cycle, making Choice *C* incorrect.

The electrons of reduced $NADP^+$ are invested in sugars in the Calvin cycle, and the electrons from reduced NAD^+ are delivered to an electron transport chain in the cristae. Neither molecule travels through ATP synthase, making Choice *A* incorrect.

3. D: Cellular respiration is the term used for the set of metabolic reactions that convert chemical bonds to energy in the form of ATP. All respiration starts with glycolysis in the cytoplasm, and in the presence of oxygen, the process will continue to the mitochondria. In a series of oxidation/reduction reactions, primarily glucose will be broken down so that the energy contained within its bonds can be transferred to the smaller ATP molecules. It's like having a $100 bill (glucose) as opposed to having one hundred $1 bills. This is beneficial to the organism because it allows energy to be distributed throughout the cell very easily in smaller packets of energy.

4. D: Allosteric activators bind to an allosteric site, a site other than the active site, of an enzyme and cause a conformational change that allows the substrate to bind to the active site of the enzyme.

5. D: An endergonic reaction absorbs energy to make bigger things from smaller things, resulting in an increase in order, or a decrease in entropy, because the molecules become condensed into a more rigid form. Choice *A* is incorrect because it does not release energy, it absorbs it. Choice *B* is incorrect because it does not increase entropy. Choice *C* is incorrect because it does not release energy and increase entropy—these are the requirements of an exergonic reaction.

Genetics

Genes

Chromosomes are found inside the nucleus of cells and contain the hereditary information of the cell in the form of **genes**. Each gene has a specific sequence of DNA that eventually encodes proteins and results in inherited traits. **Alleles** are variations of a specific gene that occur at the same location on the chromosome. For example, blue and brown are two different alleles of the gene that encodes for eye color.

Dominant and Recessive Traits

In genetics, **dominant alleles** are mostly noted in capital letters (A) and **recessive alleles** are mostly noted in lowercase letters (a). There are three possible combinations of alleles among dominant and recessive alleles: AA, Aa (known as a heterozygote), and aa. Dominant traits are phenotypes that appear when at least one dominant allele is present in the gene. Dominant alleles are considered to have stronger phenotypes and, when mixed with recessive alleles, will mask the recessive trait. The recessive trait would only appear as the phenotype when the allele combination is "aa" because a dominant allele is not present to mask it.

A gene can be pinpointed to a **locus**, or a particular position, on DNA. It is estimated that humans have approximately 20,000 to 25,000 genes. For any particular gene, a human inherits one copy from each parent for a total of two. Genotype refers to the genetic makeup of an individual within a species. Phenotype refers to the visible characteristics and observable behavior of an individual within a species.

Genotypes are written with pairs of letters that represent alleles. **Alleles** are different versions of the same gene, and, in simple systems, each gene has one dominant allele and one recessive allele. The letter of the dominant trait is capitalized, while the letter of the recessive trait is not capitalized. An individual can be homozygous dominant, homozygous recessive, or heterozygous for a particular gene. **Homozygous** means that the individual inherits two alleles of the same type while **heterozygous** means inheriting one dominant allele and one recessive allele.

If an individual has homozygous dominant alleles or heterozygous alleles, the dominant allele is expressed. If an individual has homozygous recessive alleles, the recessive allele is expressed. For example, a species of bird develops either white or black feathers. The white feathers are the dominant allele, or trait (*A*), while the black feathers are the recessive allele (*a*). Homozygous dominant (*AA*) and heterozygous (*Aa*) birds will develop white feathers. Homozygous recessive (aa) birds will develop black feathers.

Genotype (genetic makeup)	Phenotype (observable traits)
AA	white feathers
Aa	white feathers
aa	black feathers

Chromosomes

Chromosomes

In eukaryotes, chromosomes reside in the nucleus of the cell and contain genes. Although the first stage of meiosis involves the duplication of chromosomes, similar to that of mitosis, the parent cell in meiosis divides into four cells, as opposed to the two produced in mitosis.

Meiosis has the same phases as mitosis, except that they occur twice: once in meiosis I and again in meiosis II. The diploid parent has two sets of homologous chromosomes, one set from each parent. During meiosis I, each chromosome set goes through a process called **crossing over**, which jumbles up the genes on each chromatid. In anaphase one, the separated chromosomes are no longer identical and, once the chromosomes pull apart, each daughter cell is haploid (one set of chromosomes with two non-identical sister chromatids). Next, during meiosis II, the two intermediate daughter cells divide again, separating the chromatids, producing a total of four total haploid cells that each contains one set of chromosomes.

Bacterial Chromosomes and Genetic Variation

Occasionally bacteria do have DNA mutations, but they don't harbor chromosomal mutations because they only have one large, central chromosome. They don't perform recombination, instead reproducing asexually via binary fission.

However, they can still have genetic variation through the following processes:

- 1. Transformation
- 2. Transduction
- 3. Conjugation
- 4. Mutation
- 5. Transposable element relocation

Transformation	Transduction	Conjugation
Exogenous snippets of DNA enter bacterial cells.	Bacteriophages (viruses that infect bacteria) introduce foreign DNA into a bacterial cell	One bacterial cell extends a pili into another and releases DNA.

In **conjugation**, one bacterial cell is the giver (F⁺) while the other is the receiver (F⁻). After the pilus extends and attaches to the recipient, it inserts in F factor, a plasmid containing the new allele. Sometimes R plasmids, which contain and spread antibiotic-resistant genes, are transferred.

Plasmids are circular pieces of DNA that carry at least one gene. Pili actually extend from one bacterium to another, connect to the neighboring bacteria, and inject a plasmid. This process of conjugation contributes to the bacteria kingdom's diversity.

Karyotypes

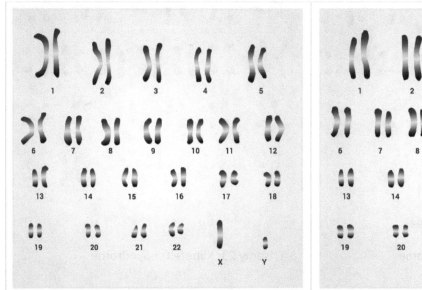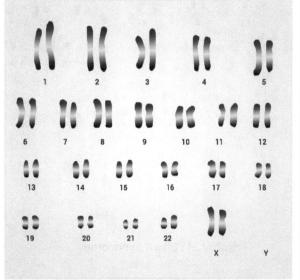

Karyotypes show a picture of an individual's 23 chromosomes and illustrate the diploid nature of our genome. Although females have two X chromosomes, only one X will be active in each cell and the other will form an inactive **Barr Body**. The inactive X chromosome is random. Some cells will have one inactive X chromosome while others have the second X chromosome inactive, resulting in an individual consisting of a mosaic of cells.

Karyotypes not only show gender, they can also illustrate occasional mistakes that occur in meiosis called **nondisjunction**. Nondisjunction results in improper separation of tetrads and chromatids. This results in one cell having an extra copy of a chromosome (**trisomy**) and another cell missing a copy of the homologous chromosome (**monosomy**). Fertilization with gametes affected by nondisjunction is often fatal, but some are viable.

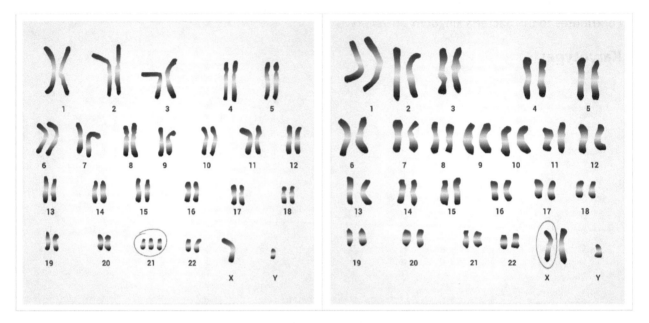

Trisomy 21: Down Syndrome   Trisomy 23: Klinefelter Syndrome

Different diseases and their inheritance patterns are listed below:

Disease	Phenotype	Cause/Pattern
Color Blindness	Red/green vision deficiency	Sex-linked recessive
Hemophilia	Blood clotting disorder	Sex-linked recessive
Muscular Dystrophy	Weak muscles and poor muscle coordination	Sex-linked recessive
Huntington's	Nervous system degeneration that has a late onset (middle age)	Autosomal dominant
Achondroplasia	Dwarfism	Autosomal dominant
Cystic Fibrosis	Excessive mucus production	Autosomal recessive
PKU	Unable to digest phenylalanine	Autosomal recessive
Tay-Sachs Disease	Intellectually disabled due to inability to metabolize lipids, death in infancy	Autosomal recessive
Sickle Cell Anemia	Red blood cell shaped like a sickle instead of a circle	Autosomal co-dominant

Disorders caused by nondisjunction and chromosomal deletions:

Disease	Phenotype	Cause/Pattern
Down Syndrome	Intellectually disabled	Trisomy 21
Klinefelter Syndrome	Sterile males	Trisomy 23 (XXY)
Turner Syndrome	Sterile females	Monosomy 21 (X)

Meiosis

Meiosis is a type of cell division in which the daughter cells have half as many sets of chromosomes as the parent cell. In addition, one parent cell produces four daughter cells. Meiosis has the same phases as mitosis, except that they occur twice—once in meiosis I and once in meiosis II. The diploid parent has two sets of chromosomes, set A and set B. During meiosis I, each chromosome set duplicates, producing a second set of A chromosomes and a second set of B chromosomes, and the cell splits into two. Each cell contains two sets of chromosomes. Next, during meiosis II, the two intermediate daughter cells divide again, producing four total haploid cells that each contain one set of chromosomes. Two of the

haploid cells each contain one chromosome of set A and the other two cells each contain one chromosome of set B.

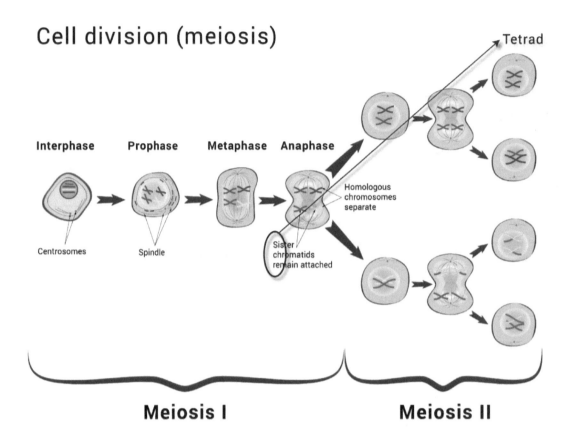

Cell division (meiosis)

Interphase   Prophase   Metaphase   Anaphase

Centrosomes   Spindle

Homologous
chromosomes
separate

Sister
chromatids
remain attached

Tetrad

Meiosis I   Meiosis II

Meiosis is similar to mitosis because it involves cellular division. However, while mitosis involves the division of somatic (body) cells, meiosis is specifically the production of gametes (egg and sperm). In mitosis, one parent cell splits once into two genetically identical and diploid daughter cells, while in meiosis, one germ cell splits twice into four genetically different, haploid daughter cells. The two divisions of meiosis (meiosis I and meiosis II) are critical because, when a sperm fertilizes an egg to create the first cell of a new organism (zygote), the zygote must have two sets of chromosomes—not four—to be viable.

The phases of meiosis I and II are nearly the same as the phases of mitosis, except for a few key differences. In mitosis, homologous chromosomes line up single file in metaphase. On the other hand, **tetrads** are paired homologous chromosomes, or homologs, which are lined up in metaphase. The tetrad is held together by the **synaptonemal complex**, which connects pairs of homologous chromosomes and is disassembled at the end of prophase I. In prophase I, tetrads additionally go through a process called **crossing over/recombination** where they exchange DNA. A **chiasmata** is where the crossing over occurs (there are 1–3 crossing over events per tetrad). Crossing over makes gametes unique because genetic recombination events are random and unpredictable. In anaphase I, the homologous chromosomes separate, moving pairs of sister chromatids (replicated chromosomes) to each side of the cell.

After meiosis I, the homologous chromosomes separate, and the two daughter cells are haploid. There is no interphase between meiosis I and meiosis II, so the DNA is not replicated. In anaphase II of meiosis, sister chromatids separate and, by the end, there are four unique, haploid daughter cells.

Although homologous chromosomes code for the same genes, there is diversity. This is because each gene has two or more alleles. **Alleles** are different forms of genes, and the **phenotype** is dependent on the **genotype**.

A process called **crossing over** occurs, which makes the daughter cells genetically different. If chromosomes didn't cross over and rearrange genes, siblings could be identical clones. There would be no genetic variation, which is a critical factor in the evolution of organisms.

Inheritance

Chromosomal Basis of Inheritance

Tenets of Mendelian genetics:

- **The law of dominance**: Dominant alleles trump recessive alleles in phenotype (the exceptions are non-Mendelian traits)

- **The law of segregation**: Alleles for each trait are separated into gametes. One allele comes from each parent, giving the offspring two copies of each allele.

- **The law of independent assortment**: Tetrads line up in metaphase I independently of other chromosomes. Each of the 23 homologues has a 50/50 chance of being on either side.

In simple Mendelian genetics, an individual can have three different genotypes. This is shown in the table below regarding the trait of flower color:

Genotype	Referred to as	Corresponding Phenotype
PP	Homozygous dominant	Purple flowers
Pp	Heterozygous	Purple flowers
pp	Homozygous recessive	White flowers

	Height	Seed Shape	Flower Color
Dominant	Tall	Round	White
Recessive Trait	Short	Wrinkled	Violet

When Mendel crossed true breeds in the P generation (as shown in the Punnett square below), he noted that all the F_1 offspring had the dominant phenotype. When he crossed F1 offspring (all heterozygotes), the F2 generation consistently showed a 3:1 phenotypic ratio of purple to white flowers. These results demonstrate the first law of Mendelian genetics: the law of dominance.

Parent (P) generation: True breed cross: PP x pp

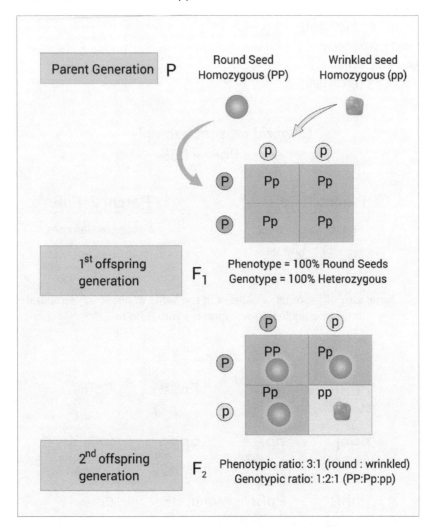

A useful application of this rule is the practice of **backcrossing**. This is when a dominant expressing organism of an unknown genotype is crossed with a recessive expressing organism. Offspring phenotype can be used to determine if the parent is homozygous or heterozygous.

Dihybrid crosses involve two traits and illustrate Mendel's law of segregation: alleles separate in meiosis.

Dihybrid crosses example: Cross $PpRr \times PpRr$

- Purple flowers = P (dominant)
- White flowers = p (recessive)
- Round seeds = R (dominant)
- Wrinkled seeds = r (recessive)

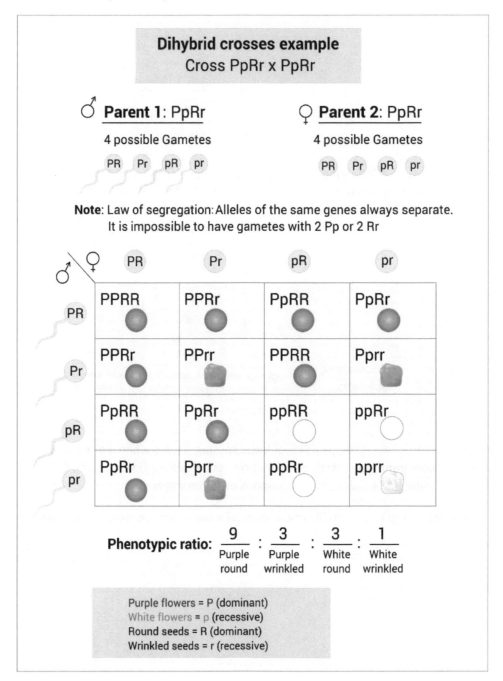

Probability can be determined by using the **law of multiplication** when all factors are present. The **law of addition** can be used to determine probability when one factor OR others may be present. Another dominant phenotype is yellow seed (Y) over green seed (y), and tall stems (T) are dominant over dwarf stems (t).

Question: What is the probability of having offspring with the genotype *PPrrYytt* if the parent cross is *PprrYyTt* × *PpRryyTt*?

Solution:

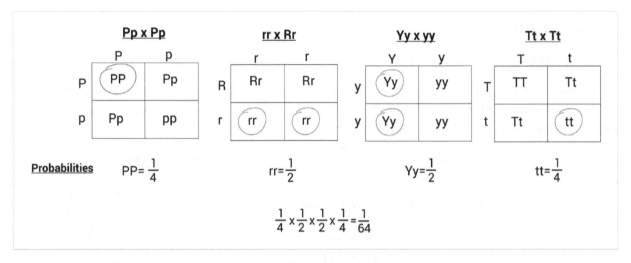

Law of Addition (Mutually Exclusive Results)

Question: What is the probability of having offspring recessive for three different traits (could contain several different combinations)?

Solution:

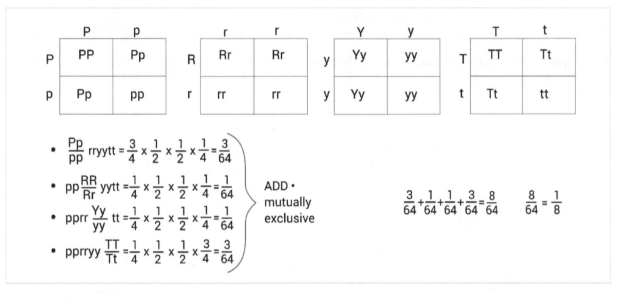

Genetic Crosses

Genetic crosses are the possible combinations of alleles, and can be represented using Punnett squares. A **monohybrid cross** refers to a cross involving only one trait. Typically, the ratio is 3:1 (DD, Dd, Dd, dd), which is the ratio of dominant gene manifestation to recessive gene manifestation. This ratio occurs when both parents have a pair of dominant and recessive genes. If one parent has a pair of dominant genes (DD) and the other has a pair of recessive (dd) genes, the recessive trait cannot be expressed in the next generation because the resulting crosses all have the Dd genotype. A **dihybrid cross** refers to one involving more than one trait, which means more combinations are possible. The ratio of genotypes for a dihybrid cross is 9:3:3:1 when the traits are not linked. The ratio for incomplete dominance is 1:2:1, which corresponds to dominant, mixed, and recessive phenotypes.

Co-Dominance and Multiple Alleles

The simple dominant/recessive model for genetics does not work for many genes.

For example, blood type is a trait that has multiple alleles: I^A, I^B, and i. I^A and I^B are "**co-dominant**" so neither is "stronger" than the other, and i is recessive to both. In the event that both co-dominant alleles are present in a genotype, both phenotypes will be present.

Genotype	Phenotype	Blood Donation Facts
IAIA, IAi	A blood (A antigens and B antibodies)	People with A blood can't receive blood from AB or B due to antibody recognition and attack of B antigen.
IAIB	AB blood (A and B antigens but no antibodies)	Universal receiver because it contains no antibodies against A or B antigens.
IBIB, IBi	B blood (B antigens and A antibodies)	People with B blood can't receive blood from AB or A due to antibody recognition and attack of A antigen.
ii	O blood (A and B antibodies)	Can only receive from other O blood (universal donor)

Blood type demonstrates the concept of co-dominance as well as multiple alleles. Below are some blood type crosses and probabilities.

Incomplete dominance occurs when the phenotype is a blending of the two alleles instead of one being dominant over the other. For example, if black and white feathers are co-dominant in birds, heterozygous offspring will have black and white speckles. However, if black and white feathers have an incomplete dominant pattern, heterozygotes will appear grey.

Sex-Linked Genetics

Normal Human Karyotype

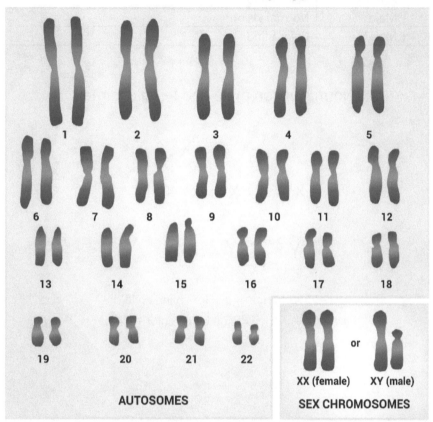

Chromosome 23 is not always homologous because it determines gender, and an individual is either XX (female) or XY (male).

The X chromosome is much larger than the Y chromosome and carries more genes, including the color-blind recessive allele. The possible genotypes, phenotypes, and inheritance patterns are shown below for the color-blind trait:

Genotype	Gender	Phenotype
XC	Female	Normal vision
XC	Female	Normal vision *She carries the allele and can pass it on to her children
Xc	Female	Color blind
XCY	Male	Normal vision
XcY	Male	Color blind

Normal vision male and female carrier

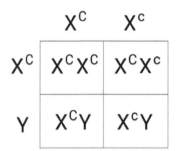

- 25% chance of a color - blind child.
- 50% chance of son being color - blind.

Genetic Modification and Biotechnology

Genes and Biotechnology

The table below lists some biotechnology tools that scientists use to study molecular genetics:

Step/Technology	Explanation
Polymerase Chain Reaction (PCR)	Primers are developed upstream and downstream of the gene of interest (engineered with deliberate restriction enzyme recognized sequences). DNA polymerase is added to the mix. There is a three-step cycle consisting of heat, cool, and warm phases which is then repeated many times. Heat separates and denatures DNA strands. Cool is where the primers anneal. Warm is when DNA polymerase elongates and synthesizes DNA.
Restriction Enzyme Digest	Restriction enzymes are a primitive bacterial defense designed to cut specific sequences of DNA. Scientists utilize them to cut open plasmids and insert their gene of interest. The same restriction enzyme is also used to cut the ends of the gene of interest. Ligase is used to glue the gene of interest into the plasmid (complementary overhanging ends anneal). *Bacterial plasmid usually carries an antibiotic resistant gene.*
Transformation	Transform bacteria with recombinant DNA. Grow bacteria on an agar plate treated with antibiotic. Only surviving clones contain the gene of interest.
Gel Electrophoresis	Separates DNA based on sequence length. DNA is negatively charged so, when placed in a gel with an electric field applied, it travels to the positive electrode and smaller segments travel faster.

Practice Questions

1. When mice develop an intermediate fur color instead of light or dark fur, what type of selection is occurring?
   a. Disruptive
   b. Stabilizing
   c. Directional
   d. Sexual

2. Blood type is a trait determined by multiple alleles, and two of them are co-dominant: I^A codes for A blood and I^B codes for B blood. The i allele codes for O blood and is recessive to both. If an A heterozygous individual and an O individual have a child, what is the probably that the child will have A blood?
   a. 25%
   b. 50%
   c. 75%
   d. 100%

Use the image below to answer questions 2–3:

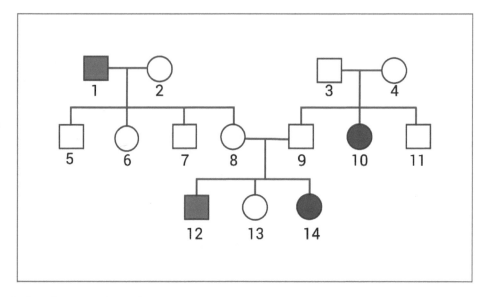

3. What kind of pedigree is shown?
   a. Autosomal dominant
   b. Autosomal recessive
   c. Sex-linked dominant
   d. Sex-linked recessive

4. What is the genotype of individual 9?
   a. *AA*
   b. *Aa*
   c. *aa*
   d. $X^A Y$

5. This karyotype indicates what about the individual?

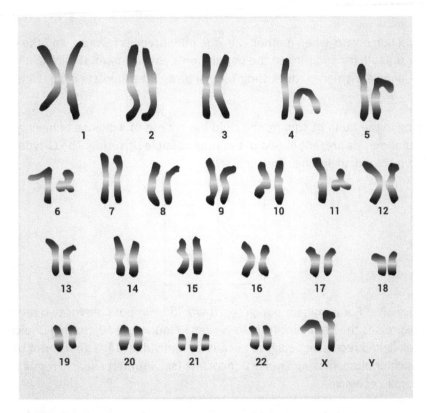

    I. They are female
    II. They are male
    III. They have Down Syndrome
    IV. They have Turner Syndrome

a. I and III
b. I and IV
c. II and III
d. II and IV

Answer Explanations

1. B: Stabilizing selection occurs when neither extreme phenotype is favored, and the intermediate phenotype is most suitable for adapting to the population's environment. If mice live in an environment with a mix of light- and dark-colored rocks, their fur will be an intermediate color. Neither light nor dark fur will be selected.

2. B: 50%. According to the Punnett square, the child has a 2 out of 4 chance of having A-type blood, since the dominant allele I^A is present in two of the four possible offspring. The O-type blood allele is masked by the A-type blood allele since it is recessive.

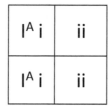

3. B: Autosomal recessive. For dominant pedigrees, it would be impossible for two recessive parents to have a child that expresses the dominant trait, as is seen in individual 10, making Choices *A* and *C* wrong. This cannot be a sex-linked recessive pedigree because of individual 14: a girl cannot be color blind unless her father is color blind, making Choice *D* incorrect (see Punnett square from #1). The correct answer is *B*, autosomal recessive.

4. B: Aa. This is an autosomal recessive pedigree, so Choice *D* is incorrect. Individual 9 has a child who has the trait, so he must have a recessive allele. He must also have the dominant allele since he does not have the trait. Choice *B* is the heterozygous genotype that has both the dominant and recessive allele, so the correct answer is *B*.

5. A: I and III. This is a female because her 23rd chromosome pair is composed of two X chromosomes and no Y. The karyotype also shows trisomy 21, which is Down syndrome. Turner syndrome is monosomy 23 (women with only one sex chromosome), making IV incorrect.

Ecology

Species, Communities, and Ecosystems

Species

Species are groups of like organisms that are able to breed with one another but not with other species.

Energy Production in Species

Organisms must consume energy, in the form of food or light, to perform anabolic reactions. If they are autotrophs, such as plants, they produce their own food. If they are heterotrophs, such as animals, they absorb the energy provided by food, most commonly sugars. The most common form of sugar used for energy is the simple sugar glucose, $C_6H_{12}O_6$ which contains extremely high amounts of potential energy stored within its atomic bonds.

The chloroplast and mitochondria are key players in energy conversions (respiration and photosynthesis) in eukaryotes. In autotrophs, the energy products created through photosynthesis are used in cellular respiration. In heterotrophs, the energy needed for cellular respiration is obtained through food. The reactants and products of each cycle are listed below:

1a. Autotrophs only: Photosynthesis in the chloroplast makes glucose and oxygen to be used in cellular respiration:

$$6CO_2 + 6H_2O \rightarrow C_6H_{12}O_6 + 6O_2$$

1b. Heterotrophs only: Consume glucose and pass its energy along the food chain

2. Both autotrophs and heterotrophs: Aerobic respiration occurs in the mitochondria, producing water, carbon dioxide, and energy.

$$C_6H_{12}O_6 + 6O_2 \rightarrow 6H_2O + 6CO_2 + 32ATP$$

3. Both autotrophs and heterotrophs: Organisms die, decompose, and their essential elements are re-used, and then the cycle repeats.

Populations

A **population** consists of all of the members of the same species that live in a geographical area and are able to reproduce and create fertile offspring. The size of any given population is determined by many factors, which include birth rate, death rate, and migration.

Population Size

The size of a population is affected by many different factors in a community. Populations grow when new members are born, and they shrink when members die or migrate away. Birth rates, death rates, and migratory rates are affected by factors such as the presence of predators, the availability of food, and the availability of shelter. The population size of predators and prey are linked. The population of a predator grows when the number of prey is great. Eventually, the population of the predator gets so high that competition exists between individuals of the predator species. The population of the prey starts to fall and the competition between the predators gets worse. Eventually, the population of the predator cannot be sustained and the population starts to fall. When the predator population decreases enough, the population of the prey starts to rebound. This relationship can also be affected by external factors. Mathematical modeling can predict this relationship and show the impact of the external factors.

Growth Curves and Carrying Capacity

Population dynamics can be characterized by **growth curves**. Growth can either be **unrestricted**, which is modeled by an exponential curve, or **restricted**, which is modeled by a logistic curve. Population growth can be restricted by environmental factors such as the availability of food and water sources, habitat, and other necessities. The **carrying capacity** of a population is the maximum population size that an environment can sustain indefinitely, given all of the above factors.

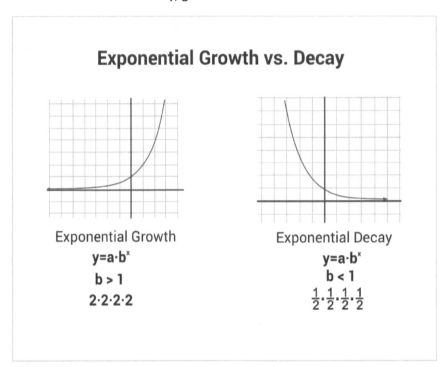

Communities

A **community** is made up of all of the different populations that live in a given geographic area. There are complex interactions both within a species and between different species. Communities that are more diverse and more complex are more stable than simple communities.

An **ecological community** is a group of species that interact and live in the same location. Because of their shared environment, they tend to have a large influence on each other.

Niche

An **ecological niche** is the role that a species plays in its environment, including how it finds its food and shelter. It could be a predator of a different species, or prey for a larger species.

Species Diversity

Species diversity is the number of different species that cohabitate in an ecological community. It has two different facets: **species richness**, which is the general number of species, and **species evenness**, which accounts for the population size of each species.

Interspecific Relationships

Different populations interact to create the complex functions of a community. These interactions can have both positive and negative impacts on the individuals involved from the different populations and can be modeled mathematically. The models can demonstrate how the negative or positive relationship will impact the population of each species in the relationship. There are five types of interactions:

Competition: Competition is when two individuals vie for a finite amount of resources, such as food, water, and mates. This can occur within or between species. Both groups are negatively affected by competition.

Predation: Predation occurs when one species, the predator, feeds on another species, the prey. Predation usually, but not always, ends in the death of the prey. This relationship is beneficial to the predator and harmful to the prey. Ultimately though, as the number of prey decrease (from predation), the number of predators eventually decrease, as food supplies dwindle.

Parasitism: Parasitism is another relationship where one species, the parasite, gains a benefit from the relationship, but the other, the host, is harmed by the relationship. Unlike predation, however, parasitism does not always result in the death of the host and does not always involve a way to ascertain food. Parasites can use their host as a food source, but they can also use their hosts as a place to lay their eggs for reproduction or to provide a habitat.

Commensalism: Commensalism is when one member of the relationship benefits and the other is not affected at all. Examples of this include when one organism uses another for transport or housing without harming the other organism.

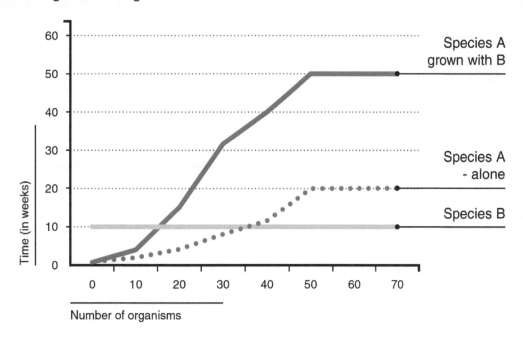

Mutualism: Mutualism is a relationship where both members benefit. There are many examples of this, including: the nectar-drinking/pollination relationship between insects or birds and plants; animals eating fruit and dispersing seeds; and animals that feed on parasites on other animals.

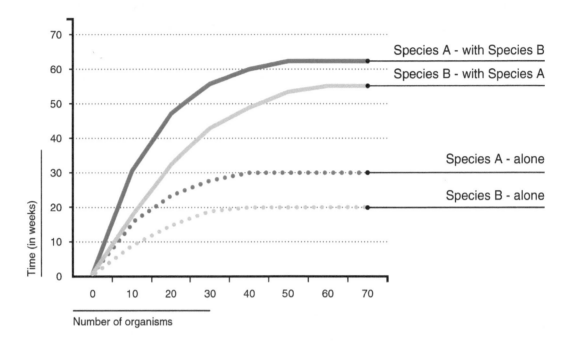

All of these relationships occur in the context of an entire ecosystem. They are not solitary relationships and are affected greatly by other forces outside of the relationship. This allows for feedback mechanisms to control the symbiotic relationships.

Population Dynamics

Population dynamics can be influenced by many factors, both biotic and abiotic. Natural and man-made disasters can greatly affect population size and species distribution. One example of this is the destruction of elm trees by Dutch elm disease. This fungus is spread by beetles and was accidentally introduced to Europe, North America, and New Zealand. The disease is native to Asia and many trees there are resistant to the disease. Dutch elm disease wiped out approximately 75% of the elm trees through much of Europe.

Ecosystems

An **ecosystem** is the basic unit of ecology and combines all of the living, or biotic, organisms and nonliving, or abiotic, components. The abiotic components include light, water, air, minerals, and nitrogen. They provide necessary nutrients and energy for the biotic components. Energy in an ecosystem usually flows from sunlight to primary producers, which are organisms that can undergo photosynthesis, to secondary consumers and finally to decomposers. Abiotic nutrients often have complex cycling systems between different members of an ecosystem.

Distribution of Ecosystems

While ecosystems are often discussed in their current state, it is important to note that ecosystems are not static; they change over time, which can be a natural process. Weather patterns, geological events, and fires are among the things that can affect an ecosystem. These items can reduce or increase

resources, destroy habitats, or even kill individuals. One example of a natural change to an ecosystem is the weather pattern known as El Niño, which is caused by increased water temperatures in the Pacific Ocean and results in weather changes throughout the world. El Niño can have an effect on many ecosystems. In the ocean off the coast of Peru, there are fewer predatory fish because the ocean conditions provide fewer nutrients for the plankton that the fish eat. In areas that experience flooding from El Niño, there can be overflow of salt water into fresh-water systems, which is destructive to those ecosystems. Areas with droughts can see a loss of producers, which is also destructive to ecosystems.

Ecosystem Diversity

Ecosystem diversity is represented by the number of different species in the ecosystem. Ecosystems with higher diversity are more resilient to changes in the environment than simple ecosystems because the loss of any one species is not as detrimental.

The key factors in maintaining diversity in an ecosystem are keystone species, producers, and essential biotic and abiotic factors. Keystone species are any species whose role in the ecosystem is disproportionate to the size of the population. Although the keystone species may not be the most numerous or the most productive part of an ecosystem, its loss would devastate the ecosystem. For example, a keystone species may be a small predator that preys on an herbivorous species and keeps that species from eliminating all of a particular plant species. If the keystone species became extinct, the herbivorous species would completely wipe out the plant species and the ecosystem would change drastically. Similarly, in large bodies of water, the sea star is a keystone species that preys on sea urchins, which helps protect the coral reefs. Within an ecosystem, each species plays a specific, important role in preserving the environment they populate together.

Biomes

A **biome** is a group of plants and animals that are found in many different continents and have the same characteristics because of the similar climates in which they live. Each biome is composed of all of the ecosystems in that area. Five primary types of biomes are aquatic, deserts, forests, grasslands, and tundra. The sum total of all biomes comprises the Earth's biosphere.

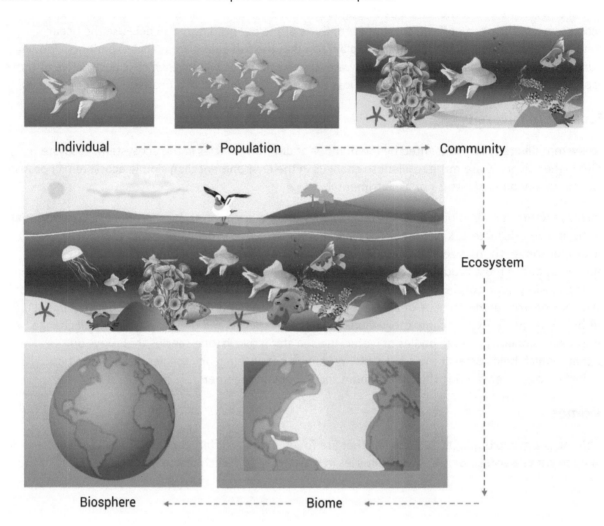

A **biosphere** is the collection of all of the biomes in the world. A biome is a group of ecosystems with similar properties, such as climate and geography.

Energy Flow

Food Webs

Energy flows from the primary producers to consumers, as consumers eat producers and other consumers. From one step to the next, 90% of the energy is lost. Therefore, in order for a tertiary consumer to receive just 1 kcal of energy, a producer needs to produce 1000 kcal of energy. The energy flow in a community can be shown as a pyramid that illustrates the loss of energy or as a web that illustrates the complex relationships between each organism.

It is important to note that the start of every single food web has a producer. Without a producer to harness the energy from the sun, the entire web would collapse. There would not be any energy to sustain any populations in the community. Producers are able to convert light energy to chemical energy via photosynthesis.

Therefore, producers (plants, protists, and even some bacteria) photosynthesize and make the food that provides energy required for all chemical reactions to occur and therefore all life to exist. A non-photosynthesizer must find and eat food, and this feeding relationship can be visualized in food chains. Consider this food chain:

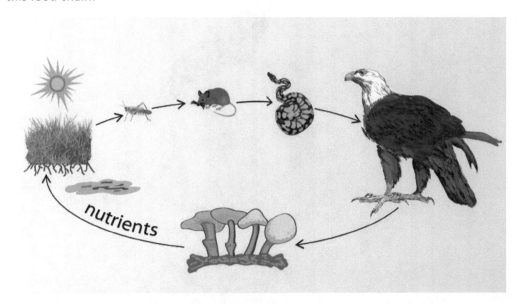

The true source of the energy for every living organism is the sun. Plants absorb the sun's energy to make glucose and are on the first trophic level (feeding level). The grasshopper on the second trophic level is an example of an herbivore and is a primary consumer, as he is the first eater in the food chain. Unfortunately, he receives only 10 percent of the energy that the plant absorbed (this is known as the 10 percent rule) because the other 90 percent of energy was either used by the plant to grow or will be lost as heat. The mouse on the third trophic level is the secondary consumer, or second eater. Food chains are not as inclusive as food webs, which show all feeding relationships in an ecosystem. Looking at this food chain suggests that mice are carnivores (eaters of animals), but mice also eat berries and plants, so they are actually considered omnivores (eaters of both plants and animals). The mouse only gets 10 percent of the energy from the grasshopper, which is actually only 1 percent of the original energy provided by the Sun. The snake on the fourth trophic level is a carnivore, as is the hawk on the highest trophic level.

The arrows in the food chain show the transfer of energy, and fungi as well as bacteria act as decomposers, which break down organic material. Decomposers act at every trophic level because they feed on all organisms; they are non-discriminating omnivores. Decomposers are critical for life, as they recycle the atoms and building blocks of organisms.

Carbon Cycling

Carbon cycle: Carbon forms the backbone of all biologically important molecules. It is found in the atmosphere as CO_2. Plants, algae, and cyanobacteria take CO_2 and make carbohydrates during photosynthesis using energy from the sun. The carbon then moves through animals in the food chain and is returned to the atmosphere as CO_s during respiration. Decaying biological material, called **detritus**, also provides carbon to the soil. A final source of carbon is the burning of wood and fossil fuels, which releases CO_2 into the atmosphere.

The Carbon Cycle

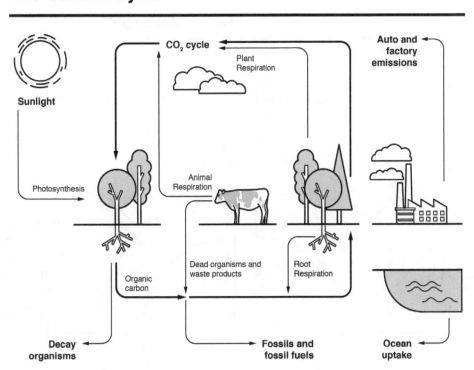

Climate Change

Climate Change and Greenhouse Gases

Greenhouse gases in the Earth's atmosphere include water vapor, carbon dioxide, methane, nitrous oxide, and chlorofluorocarbons (CFCs), which trap heat between the surface of the Earth and the Earth's lowest atmospheric layer, the troposphere. The increase of these gases leads to warming or cooling trends that cause unpredictable or unprecedented meteorological shifts. These shifts can cause natural disasters, affect plant and animal life, and dramatically impact human health. Water vapor is a naturally found gas, but as the Earth's temperature rises, the presence of water vapor increases; as water vapor increases, the Earth's temperature rises. This creates a somewhat undesirable loop. Carbon dioxide is produced through natural causes, such as volcanic eruptions, but also is greatly affected by human activities, such as burning fossil fuels. A significant increase in the presence of atmospheric carbon dioxide has been noted since the Industrial Revolution; this is important as carbon dioxide is considered the most significant influencer of climate change on Earth. Methane is produced primarily from animal and agriculture waste and landfill waste. Nitrous oxide is primarily produced from the use of fertilizers and fossil fuels. CFCs are completely synthetic and were previously commonly found in aerosol and other high-pressure containers; however, after being linked to ozone layer depletion, they have been stringently regulated internationally and are now in limited use. Scientists have stated that the climate shifts recorded since the Industrial Revolution cannot be attributed to natural causes alone, as the patterns do not follow those of climate shifts that took place prior to the Industrial Revolution.

Natural Greenhouse Effect vs. Human Influence

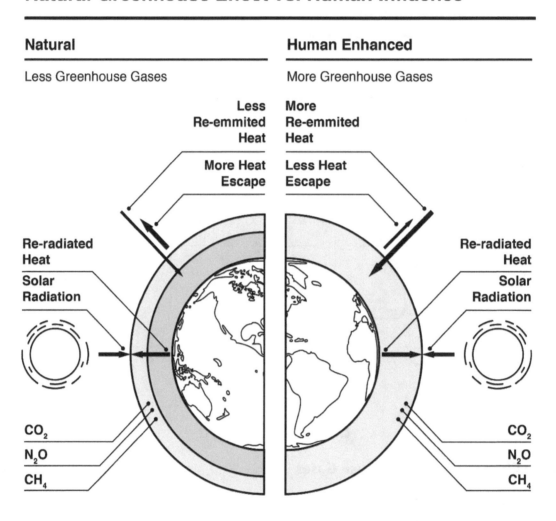

Practice Questions

1. Two different plants were grown in a lab and their response to light was investigated. Based on the representative qualitative data shown in the diagram, which of these mechanisms would best explain the flowering patterns?

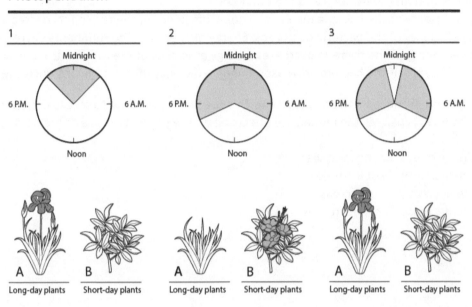

Photoperiodism

    a. Auxin's property of elongating cells exposed to light is responsible for flowering in both plants.
    b. Plant B has a selective advantage because it is unusually reproductively active in the winter, which reduces competition.
    c. Plant A requires a threshold of sunlight in order to flower. Its blooming is independent of the season.
    d. Both plants are dependent on the amount of continuous light exposure in order to flower.

2. A characteristic of life is that organisms are able to react and respond to their environment. The choices below are proposed physiological regulatory responses of various organisms. Which one actually occurs?
    a. Fungi rely on phototropism to aid in obtaining resources.
    b. Humans regulate temperature solely by sweating and shivering.
    c. Plants maintain circadian rhythms, such as timing when their stomata open.
    d. Protists use the hormone endocrine-driven feedback loops to regulate processes.

3. What happens to the population of a predator if the population of a prey decreases?
    a. Increases
    b. Decreases
    c. Stay the same
    d. Not enough information to answer the question

4. Certain bacteria that can break down the bonds in cellulose live in the gut of ruminants, which are mammals that feed primarily on grasses. Animals cannot break down cellulose. How does this affect the energy efficiency of both the bacteria and the ruminants?
    a. Energy efficiency of the bacteria increases. Energy efficiency of the ruminants decreases.
    b. Energy efficiency of the bacteria decreases. Energy efficiency of the ruminants decreases.
    c. Energy efficiency of the bacteria decreases. Energy efficiency of the ruminants increases.
    d. Energy efficiency of the bacteria increases. Energy efficiency of the ruminants increases.

5. A new species is introduced into an ecosystem. This species is a parasite to an existing species in the ecosystem, the host. What will the immediate effects be on the population size of the parasite and the host?
    a. Parasite increases, host decreases
    b. Parasite increases, host increases
    c. Parasite decreases, host increases
    d. Parasite decreases, host decreases

Answer Explanations

1. B: Photoperiodism is the phenomenon where dark exposure determines flowering, and therefore Choices *C* and *D* are wrong. Choice *B* is correct because plant B is dependent on the amount of darkness—not on the amount of light. The interrupted darkness in scenario three prevents flowering of the plant. Flowering in the winter, or when the amount of light is limited, is unusual because usually springtime is when flowers make their appearance. This could be considered an adaptation because it reduces competition for pollinators, since other flowers are not present to compete, ensuring the plant passes on its genes. Choice *A* is untrue because auxin is responsible for stem elongation and is not related to flowering.

2. C: Fungi are not photosynthetic, making Choice *A* incorrect. Choice *B* is incorrect because temperature regulation is more complicated than that. For example, induction of fever as part of the immune system's response to pathogens is under the umbrella of temperature regulation. Choice *D* is incorrect because protists are single-celled organisms for the most part, and the multi-cellular ones are not differentiated enough to have organs; therefore, they do not have an endocrine system.

3. B: When the population of a prey decreases, the population of the predator will also decrease as competition increases between the individuals in the predator population and as the prey resource becomes scarce.

4. D: The ruminants provide a food source for the bacteria, and the bacteria help the ruminants utilize their main food source. Therefore, the energy efficiency of both organisms increases.

5. A: A parasite benefits from the relationship with the host, while the host suffers a fitness cost. Therefore, the population of the parasite will increase while the population of the host will decrease.

Evolution and Biodiversity

Evidence for Evolution

The Fossil Record

Fossils are the preserved remains of animals and organisms from the distant past. They provide evidence of evolution and can elucidate the homology of both living and extinct species. Looking at the **fossil record** over time can help identify how quickly or slowly evolutionary changes occurred, and can also help match those changes to environmental changes that were occurring concurrently.

Comparative Genetics

In **comparative genetics**, different organisms are compared at a genetic level to look for similarities and differences. DNA sequence, genes, gene order, and other structural features are among the features that may be analyzed in order to look for evolutionary relationships and common ancestors between the organisms. Comparative genetics was useful in elucidating the similarities between humans and chimpanzees and linking their evolutionary history.

Homology

Organisms that developed from a common ancestor often have similar characteristics that function differently. This similarity is known as **homology**. For example, humans, cats, whales, and bats all have bones arranged in the same manner from their shoulders to their digits. However, the bones form arms in humans, forelegs in cats, flippers in whales, and wings in bats, and these forelimbs are used for lifting, walking, swimming, and flying, respectively. The similarity of the bone structure shows a common ancestry, but the functional differences are the product of evolution.

Homologous Structures

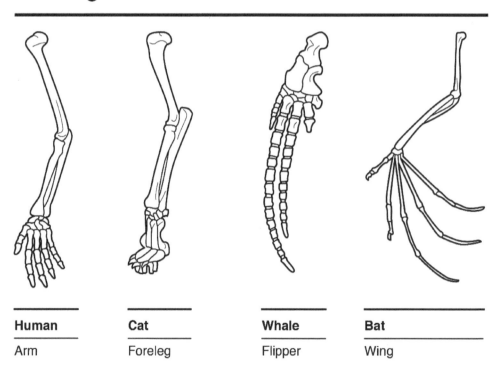

Human	Cat	Whale	Bat
Arm	Foreleg	Flipper	Wing

Natural Selection

Evolution is the concept that there is one common ancestor for all living organisms, and, over time, genetic variation and mutations cause the development of different species. According to this concept, populations of organisms evolve, not individuals, and over time, genetic variation and mutations lead to such changes. Charles Darwin came up with a scientific model of evolution based on the idea that individuals within a population can have longer lives (better survival) and higher reproduction rates based on certain specific traits that they have inherited, called **natural selection**. The variation of a trait that enhances survival and reproduction in the environment is the one that gets passed on. The survival and inheritance of these traits through many subsequent generations causes a change in the overall population. The traits that are more advantageous for survival and reproduction become more common in subsequent generations and increase the diversity of the population. For example, when there was a drought in the Galapagos Islands, the finches with large beaks became more populous because they were able to survive on the larger, rougher seeds that were remaining.

Evolutionary Fitness

Sexual selection is a type of natural selection in which individuals with certain traits are more likely to find a mate than individuals without those traits. This can occur through direct competition of one sex for a mate of the opposite sex. For example, larger males may prevent smaller males from mating by using their size advantage to keep them away from the females. Sexual selection can also occur through mate choice. This can happen when individuals of one sex are choosy about their mate of the opposite sex, often judging their potential mate based on appearance or behavior. For example, female peacocks often mate with the showiest male with large, beautiful feathers. In both types of sexual selection, individuals with some traits have better reproductive success, and the genes for those traits become more prevalent in subsequent populations.

Adaptations are Favored by Natural Selection

Adaptations are inherited characteristics that enhance survival and reproductive capabilities in specific environments. Charles Darwin's idea of natural selection explains *how* populations change—adaption explains *why*. Darwin based his concept of evolution on three observations: the unity of life, the diversity of life, and the suitability of organisms for their environments. There was unity in life based on the idea that all organisms descended from common ancestors. Then, as the descendants of the common ancestors faced changes in their environments, they moved to new environments. There they adapted new features to help them in their new way of life. This concept explains the diversity of life and how organisms are matched to their environments.

An example of natural selection is found in penguins—birds that cannot fly. Over time, populations of penguins lost the ability to fly but became master swimmers. Their habitats are surrounded by water, and their food sources are in the water. Penguins that could dive for food survived better than those that could fly, and the divers produced more offspring. The gene pool changed as a result of natural selection.

Populations in Hardy-Weinberg Equilibrium

All populations have genetic diversity, but some populations aren't changing. The **gene pool** consists of all copies of every allele at every locus in every member of a population. If the allele and genotype frequencies of a population don't change between generations, the population is in a **Hardy-Weinberg**

(HW) equilibrium, named for the British mathematician and German physician who came up with the concept in 1908. There are five conditions that must be met for a HW equilibrium: (1) a large population size, (2) absence of migration, (3) no net mutations, (4) random mating, and (5) absence of selection.

The HW equation calculates the frequency of phenotypes in a population that isn't evolving and is written as follows: $p^2 + 2pq + q^2 = 1$, where p is the frequency of one allele, q is the frequency of the other allele, and pq is the frequency of the alleles mixing. P and q must add up to equal 1. As in the figure below, in a given population of wildflowers, the frequency of the red flower allele (p) is 80%, and the frequency of the white flower allele (q) is 20%. Therefore, $p = 0.8$ and $q = 0.2$. In a non-evolving population, the frequency of red flowers would be $p^2 = 0.64 = 64\%$, the frequency of pink flowers as a mix of red flower and white flower alleles would be $2pq = 0.32 = 32\%$, and the frequency of white flowers would be $q^2 = 0.04 = 4\%$. If the frequency of any flower color doesn't match the calculations from the HW equation, then the population is evolving.

Parameters for Natural Selection

There are three important points to remember about natural selection. Although natural selection occurs due to an individual organism's relationship to its environment, it is a population—not individuals— that change over time. Second, natural selection only affects heritable traits that vary within a population. If all individuals within a population share an identical trait, natural selection cannot occur, and that trait will not be modified. Lastly, which traits are the favored traits is always changing. The environment is an important factor in natural selection, so if the environment changes, a trait that was previously favored may no longer be beneficial. Natural selection is a fluid process that is always at work.

Speciation and Isolation Methods

Speciation is the method by which one species splits into two or more species due to either geographic separation, called allopatric speciation, or a reduction in gene flow between varying members of the population, called sympatric speciation. In **allopatric speciation**, one population is divided into two subpopulations. For example, if a drought occurs and a large lake becomes divided into two smaller lakes, each lake is left with its own population that cannot intermingle with the population of the other lake. When the genes of these two subpopulations are no longer mixing with each other, new mutations can arise and natural selection can take place.

In **sympatric speciation**, gene flow in the population is reduced by polyploidy, sexual selection, and habitat differentiation. **Polyploidy** is more common in plants than animals and results when cell division during reproduction creates an extra set of chromosomes. In **sexual selection**, organisms of one sex choose their mate of the opposite sex based on certain traits. If there is high selection for two extreme variations of a trait, sympatric speciation may occur. **Habitat differentiation** occurs when a subpopulation exploits a resource that is not used by the parent population. Both allopatric and sympatric speciation can occur quickly or slowly, and may involve just a few gene changes or many gene changes between the new species.

One important distinguishing factor in the formation of two species is their **reproductive isolation**. Species are characterized by their members' ability to breed and produce viable offspring. When speciation occurs and new species are formed, there must have been a biological barrier that prevented the two species from producing viable offspring.

Following speciation, there are two types of **reproductive barriers** that keep the two populations from mating with each other. These are classified as either prezygotic barriers or postzygotic barriers. **Prezygotic barriers** prevent fertilization via habitat isolation, temporal isolation, and behavioral isolation. Through habitat isolation, two species may inhabit the same area but don't often encounter each other. **Temporal isolation** is when species breed at different times of the day, during different seasons, or during different years, so their mating patterns never coincide. **Behavioral isolation** refers to mating rituals that prevent an organism from recognizing a different species as potential mate.

Other prezygotic barriers block fertilization after a mating attempt. **Mechanical isolation** occurs when anatomical differences prevent fertilization. **Gametic isolation** occurs when the gametes of two species are incompatible.

Environmental Change Serve as Selective Mechanisms

The environment constantly changes, which drives selection. Although an individual's traits are determined by their genotype, or makeup of genes, natural selection more directly influences phenotype, or observable characteristics. The outward appearance or ability of individuals affects their ability to adapt to their environment and survive and reproduce. Phenotypic changes occurring in a population over time are accompanied by changes in the gene pool.

The classic example of this is the peppered moth. It was once a light-colored moth with black spots, though a few members of the species had a genetic variation resulting in a dark color. When the Industrial Revolution hit London, the air became filled with soot and turned the white trees darker in color. Birds were then able to spot and eat the light-colored moths more easily. Within just a few months, the moths with genes for darker color were better able to avoid predation. Subsequent generations had far more dark-colored moths than light ones. Once the Industrial Revolution ended and the air cleared, light-colored moths were better able to survive, and their numbers increased.

Causes of Phenotypic Variations

There are three ways in which phenotypes change over time due to natural selection: directional selection, disruptive selection, and stabilizing selection. **Directional selection** occurs when an extreme phenotypic variation is favored. This generally happens when a population's environment changes or the population migrates to a new habitat. When the Galapagos Islands suffered a drought, finches with larger beaks were able to eat the larger, tougher seeds that became abundant. Thus, finches with that phenotype survived and reproduced more often, and that trait became more prevalent in subsequent generations. **Disruptive selection** occurs when both extremes of a phenotype are favored. Finches in Cameroon have either large beaks or small beaks. The large-beaked birds are efficient at eating large, tough seeds; the small-beaked birds are adept at eating small seeds. Birds with medium-sized beaks were not adept at eating either size of seed, so selection favored the other finches. **Stabilizing selection** occurs when neither extreme phenotype is favored, and the intermediate phenotype is best suited for adapting to the population's environment. If mice live in an environment with a mix of light and dark colored rocks, mice with an intermediate fur color are favored. Neither light nor dark fur will be selected.

The Effect of Phenotypic Variations on Fitness

Geneticists have a specialized definition of **fitness**. They use the term to denote an organism's capacity to survive, mate, and reproduce. This ultimately equates to the probability or likelihood that the organism will be able to pass on its genetic information to the next generation. Fitness does not mean the strongest, biggest, or most dangerous individual. A more subtle combination of anatomy, physiology, biochemistry, and behavior determine genetic fitness.

Another way to understand genetic fitness is by knowing that phenotypes affect survival and the ability to successfully reproduce. Phenotypes are genetically determined and genes contributing to fitness tend to increase over time.

Thus, the "fittest" organisms survive and pass on their genetic makeup to the next generation. This is what Darwin meant by "survival of the fittest," which is the cornerstone of Darwin's theory of evolution.

Influence of Phenotype on Genotype

Natural selection provides processes that tend to increase a population's adaptive abilities. The strongest phenotypes survive, prosper, and pass on their genetic code to the next generation. The new generation of phenotypes has a fitter genotype because they have inherited more adaptive characteristics.

Thus, the fit thrive while the weak become extinct over time. This pattern, when repeated over many generations, develops strong, fit phenotypes that survive and reproduce offspring who are as fit as or fitter than their parents. Sometimes this apparently inexorable movement can be modified by drastic external conditions such as wide variations in climate. It is important to remember that all extinct species were once fit and adapted to their environments. Unforeseen circumstances always have the potential to cause chaos in the physical world.

Other Causes of Genetic Changes in Species and Populations

While the concept of natural selection focuses on changes over time in relation to the environment, there are other circumstances when change occurs randomly. Not all genetic changes relate to survival and reproduction. **Genetic drift** is the idea that the alleles of a gene can change unpredictably between generations due to chance events. If certain alleles are lost between generations, the genetic diversity of the population decreases because that genetic variation is lost forever. For example, a population of wildflowers consists of red flowers (RR and Rr) and white flowers (rr). If a large animal destroys all of the white wildflowers, the subsequent generation could be left with no (or far fewer) alleles for white flowers. Genetic drift has the greatest effect on smaller populations. Certain alleles can be over- or under-represented, even if they are not advantageous. In addition, harmful alleles can become fixed if their normal counterpart becomes extinct.

A **population bottleneck** is a type of genetic drift. This occurs when a population significantly decreases, usually due to a sudden change in the environment such as a flood or a fire. In the surviving population, certain alleles may be over- or under-represented, and others may be completely missing. Even if the surviving population returns to its original size, it will lack the genetic diversity of the original population. The **founder effect** is a special case of the bottleneck effect. It occurs when a few individuals become separated from the larger population and form their own new population. The frequency of non-dominant alleles may increase in the new smaller population, as may the frequency of inherited disorders due to a lack of dominant alleles that would keep the disorder inactive.

Classification of Biodiversity

Classification Schemes

Taxonomy is the science behind the biological names of organisms. Biologists often refer to organisms by their Latin scientific names to avoid confusion with common names, such as with fish. Jellyfish, crayfish, and silverfish all have the word "fish" in their name, but belong to three different species. In the eighteenth century, Carl Linnaeus invented a naming system for species that included using the Latin scientific name of a species, called the **binomial**, which has two parts: the **genus**, which comes first, and the **specific epithet**, which comes second. Similar species are grouped into the same genus. The Linnaean system is the commonly used taxonomic system today and, moving from comprehensive similarities to more general similarities, classifies organisms into their species, genus, family, order, class, phylum, and kingdom. *Homo sapiens* is the Latin scientific name for humans.

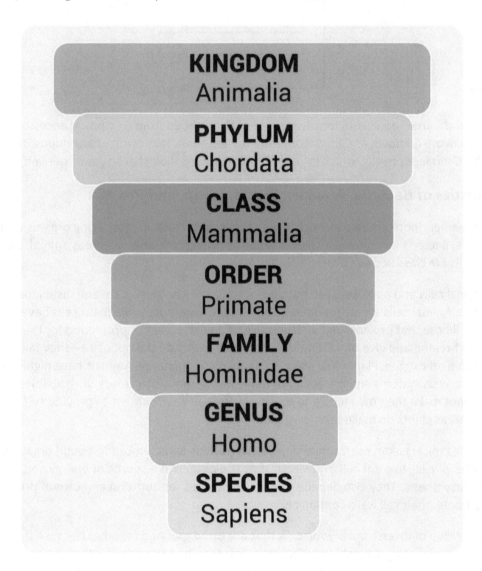

Phylogenetic trees are branching diagrams that represent the evolutionary history of a species. The branch points most often match the classification groups set forth by the Linnaean system. Using this system helps elucidate the relationship between different groups of organisms. The diagram below is that of an empty phylogenetic tree:

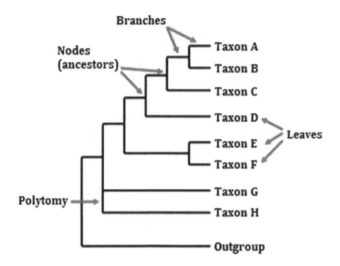

Each branch of the tree represents the divergence of two species from a common ancestor. For example, the coyote is known as Canis latrans and the gray wolf is known as Canis lupus. Their common ancestor, the Canis lepophagus, which is now extinct, is where their shared genus derived.

Characteristics of Bacteria, Animals, Plants, Fungi, and Protists

As discussed earlier, there are two distinct types of cells that make up most living organisms: prokaryotic and eukaryotic. Bacteria (and archaea) are classified as prokaryotic cells, whereas animal, plant, fungi, and protist cells are classified as eukaryotic cells.

Although animal cells and plant cells are both eukaryotic, they each have several distinguishing characteristics. Animal cells are surrounded by a plasma membrane, while plant cells have a cell wall made up of cellulose that provides more structure and an extra layer of protection for the cell. Animals use oxygen to breathe and give off carbon dioxide, while plants do the opposite—they take in carbon dioxide and give off oxygen. Plants also use light as a source of energy. Animals have highly developed sensory and nervous systems and the ability to move freely, while plants lack both abilities. Animals, however, cannot make their own food and must rely on their environment to provide sufficient nutrition, whereas plants do make their own food.

Fungal cells are typical eukaryotes, containing both a nucleus and membrane-bound organelles. They have a cell wall, similar to plant cells; however, they use oxygen as a source of energy and cannot perform photosynthesis. They also depend on outside sources for nutrition and cannot produce their own food. Of note, their cell walls contain chitin.

Protists are a group of diverse eukaryotic cells that are often grouped together because they do not fit into the categories of animal, plant, or fungal cells. They can be categorized into three broad categories: protozoa, protophyta, and molds. These three broad categories are essentially "animal-like," "plant-like," and "fungus-like," respectively. All of them are unicellular and do not form tissues. Besides this

simple similarity, protists are a diverse group of organisms with different characteristics, life cycles, and cellular structures.

Cladistics

Phylogenetics is the subfield of biology that studies the evolutionary history of a species or group of species and their relationships. Heritable traits are evaluated to make conclusions about similarities and differences between organisms. Analyzing these characteristics increases our understanding of species and population evolution.

Cladistics is a method of classifying organisms based primarily on their proposed common ancestry. Using this method, species are grouped into **clades**, which include one ancestral species and all of its descendants. Some assume that species with similar traits are related. However, these similarities may appear by analogy, which means that the species were subject to a similar natural selection process but don't share a common ancestor. Cladograms help discern the difference between analogous features and homologous features. Below is an example of a cladogram:

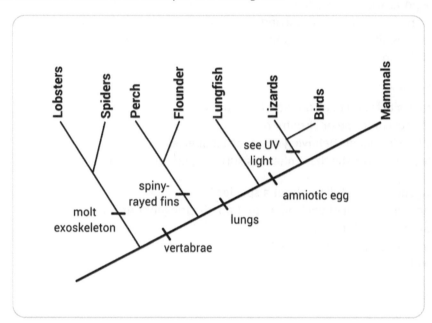

While a phylogenetic tree diagrams an organism's believed evolutionary history, a cladogram specifies the characteristics that change within the descendent groups, making it easy to see the homology of traits between related species.

Practice Questions

1. Charles Darwin's theory of evolution is based on what type of selection?
   a. Natural selection
   b. Sexual selection
   c. Disruptive selection
   d. Stabilizing selection

2. The founder effect occurs when which of the following occur?
   a. A new species suddenly fills an open niche.
   b. A new species is developed.
   c. Individuals develop an extreme phenotype through natural selection.
   d. A few individuals become separated from the larger population and form a new population.

3. What do evolutionary theorists believe the Hardy-Weinberg equation tells us?
   a. Whether a population is evolving
   b. The type of natural selection occurring
   c. If genetic drift is altering a population
   d. The size of the population

4. What is an adaptation?
   a. The original traits found in a common ancestor
   b. Changes that occur in the environment
   c. When one species begins behaving like another species
   d. An inherited characteristic that enhances survival and reproduction

5. What do phylogenetic trees tell us about a species?
   a. The genetic contribution of each allele that an offspring inherits
   b. Their proposed evolutionary history
   c. How many alleles exist for a specific trait of the species
   d. Their eye color

Answer Explanations

1. A: Charles Darwin founded the theory of natural selection. He believed that stronger individuals would continue to thrive while weaker individuals would die off.

2. D: The founder effect occurs when a few individuals from a population become separated from the larger population and form their own new population. This may occur when a storm blows a few individuals to a new island or habitat. The frequency of non-dominant alleles may increase in the new smaller population, as well as the frequency of inherited disorders due to a lack of dominant alleles that would keep the disorder inactive.

3. A: All populations have genetic diversity, but according to evolutionists, that doesn't guarantee the population is evolving. In order to assess whether a population is evolving, scientists use a mathematical equation to calculate the phenotypes of a non-evolving population. The results of that equation can be compared to the actual phenotypes seen in the population. If the allele and genotype frequencies of a population aren't changing between generations, the population is described as being in a Hardy-Weinberg (HW) equilibrium.

4. D: Charles Darwin based the idea of adaptation around his original concept of natural selection. He believed that evolution occurred based on three observations: the unity of life, the diversity of life, and the suitability of organisms to their environments. There was unity in life based on the idea that all organisms descended from a common ancestor. Then, as the descendants of common ancestors faced changes in their environments or moved to new environments, they began adapting new features to help them. This concept explained the diversity of life and how organisms were matched to their environments. Natural selection helps to improve the fit between organisms and their environments by increasing the frequency of features that enhance survival and reproduction.

5. B: Phylogenetic trees are used to illustrate the believed evolutionary history of a species. They are branching diagrams, and the branch points most often match the classification groups set forth by the Linnaean system. Using this system helps elucidate the relationship between different groups of organisms.

Digestion and Absorption

Digestive System

The human body relies completely on the **digestive system** to meet its nutritional needs. After food and drink are ingested, the digestive system breaks them down into their component nutrients and absorbs them so that the circulatory system can transport them to other cells to use for growth, energy, and cell repair. These nutrients may be classified as proteins, lipids, carbohydrates, vitamins, and minerals.

The digestive system is thought of chiefly in two parts: the digestive tract (also called the alimentary tract or gastrointestinal tract) and the accessory digestive organs. The **digestive tract** is the pathway in which food is ingested, digested, absorbed, and excreted. It is composed of the mouth, pharynx, esophagus, stomach, small and large intestines, rectum, and anus. **Peristalsis**, or wave-like contractions of smooth muscle, moves food and wastes through the digestive tract. The accessory digestive organs are the salivary glands, liver, gallbladder, and pancreas.

Mouth and Stomach

The mouth is the entrance to the digestive system. Here, the mechanical and chemical digestion of the food begins. The food is chewed mechanically by the teeth and shaped into a **bolus** by the tongue so that it can be more easily swallowed by the esophagus. The food also becomes more watery and more pliable with the addition of saliva secreted from the salivary glands, the largest of which are the parotid glands. The glands also secrete **amylase** in the saliva, an enzyme which begins chemical digestion and breakdown of the carbohydrates and sugars in the food.

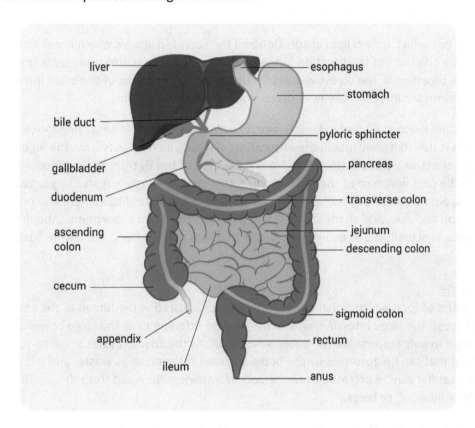

The food then moves through the pharynx and down the muscular esophagus to the stomach.

The stomach is a large, muscular sac-like organ at the distal end of the esophagus. Here, the bolus is subjected to more mechanical and chemical digestion. As it passes through the stomach, it is physically squeezed and crushed while additional secretions turn it into a watery nutrient-filled liquid that exits into the small intestine as **chyme**.

The stomach secretes a great many substances into the **lumen** of the digestive tract. Some cells produce gastrin, a hormone that prompts other cells in the stomach to secrete a gastric acid composed mostly of hydrochloric acid (HCl). The HCl is at such a high concentration and low pH that it denatures most proteins and degrades a lot of organic matter. The stomach also secretes mucous to form a protective film that keeps the corrosive acid from dissolving its own cells. Gaps in this mucous layer can lead to peptic ulcers. Finally, the stomach also uses digestive enzymes like proteases and lipases to break down proteins and fats; although there are some gastric lipases here, the stomach is most known for breaking down proteins.

Small Intestine

The chyme from the stomach enters the first part of the small intestine, the **duodenum**, through the **pyloric sphincter**, and its extreme acidity is partly neutralized by sodium bicarbonate secreted along with mucous. The presence of chyme in the duodenum triggers the secretion of the hormones secretin and cholecystokinin (CCK). Secretin acts on the pancreas to dump more sodium bicarbonate into the small intestine so that the pH is kept at a reasonable level, while CCK acts on the gallbladder to release the **bile** that it has been storing. Bile is a substance produced by the liver and stored in the gallbladder which helps to **emulsify** or dissolve fats and lipids.

Because of the bile which aids in lipid absorption and the secreted lipases which break down fats, the duodenum is the chief site of fat digestion in the body. The duodenum also represents the last major site of chemical digestion in the digestive tract, as the other two sections of the small intestine (the **jejunum** and **ileum**) are instead heavily involved in absorption of nutrients.

The small intestine reaches 40 feet in length, and its cells are arranged in small finger-like projections called **villi**. This is due to its key role in the absorption of nearly all nutrients from the ingested and digested food, effectively transferring them from the lumen of the GI tract to the bloodstream where they travel to the cells which need them. These nutrients include simple sugars like glucose from carbohydrates, amino acids from proteins, emulsified fats, electrolytes like sodium and potassium, minerals like iron and zinc, and vitamins like D and B12. Vitamin B12's absorption, though it takes place in the intestines, is actually aided by **intrinsic factor** that was released into the chyme back in the stomach.

Large Intestine

The leftover parts of food which remain unabsorbed or undigested in the lumen of the small intestine next travel through the **large intestine**, which may also be referred to as the large bowel or colon. The large intestine is mainly responsible for water absorption. As the chyme at this stage no longer has anything useful that can be absorbed by the body, it is now referred to as **waste**, and it is stored in the large intestine until it can be excreted from the body. Removing the liquid from the waste transforms it from liquid to solid stool, or **feces**.

This waste first passes from the small intestine to the **cecum**, a pouch which forms the first part of the large intestine. In herbivores, it provides a place for bacteria to digest cellulose, but in humans most of it is vestigial and is known as the appendix. From the cecum, waste next travels up the ascending colon, across the transverse colon, down the descending colon, and through the sigmoid colon to the rectum. The rectum is responsible for the final storage of waste before being expelled through the **anus**. The anal canal is a small portion of the rectum leading through to the anus and the outside of the body.

Pancreas

The **pancreas** has endocrine and exocrine functions. The endocrine function involves releasing the hormones insulin, which decreases blood sugar (glucose) levels, and glucagon, which increases blood sugar (glucose) levels, directly into the bloodstream. Both hormones are produced in the **islets of Langerhans**, insulin in the beta cells and glucagon in the alpha cells.

The major part of the gland has exocrine function, which consists of acinar cells secreting inactive digestive enzymes (**zymogens**) into the main pancreatic duct. The main pancreatic duct joins the common bile duct, which empties into the small intestine (specifically the duodenum). The digestive enzymes are then activated and take part in the digestion of carbohydrates, proteins, and fats within chyme (the mixture of partially digested food and digestive juices).

The Blood System

Circulatory System

The **cardiovascular system** (also called the circulatory system) is a network of organs and tubes that transport blood, hormones, nutrients, oxygen, and other gases to cells and tissues throughout the body. It is also known as the cardiovascular system. The major components of the circulatory system are the blood vessels, blood, and heart.

Blood Vessels

In the circulatory system, **blood vessels** are responsible for transporting blood throughout the body. The three major types of blood vessels in the circulatory system are arteries, veins, and capillaries. **Arteries** carry blood from the heart to the rest of the body. **Veins** carry blood from the body to the heart. **Capillaries** connect arteries to veins and form networks that exchange materials between the blood and the cells.

In general, arteries are stronger and thicker than veins, as they withstand high pressures exerted by the blood as the heart pumps it through the body. Arteries control blood flow through either **vasoconstriction** (narrowing of the blood vessel's diameter) or **vasodilation** (widening of the blood vessel's diameter). The smallest arteries, which are farthest from the heart, are called **arterioles**. The blood in veins is under much lower pressures, so veins have valves to prevent the backflow of blood.

Most of the exchange between the blood and tissues takes place through the capillaries. There are three types of capillaries: continuous, fenestrated, and sinusoidal.

Continuous capillaries are made up of epithelial cells tightly connected together. As a result, they limit the types of materials that pass into and out of the blood. Continuous capillaries are the most common type of capillary. **Fenestrated capillaries** have openings that allow materials to be freely exchanged between the blood and tissues. They are commonly found in the digestive, endocrine, and urinary systems. **Sinusoidal capillaries** have larger openings and allow proteins and blood cells through. They are found primarily in the liver, bone marrow, and spleen.

Blood

Blood is vital to the human body. It is a liquid connective tissue that serves as a transport system for supplying cells with nutrients and carrying away their wastes. The average adult human has five to six quarts of blood circulating through their body. Approximately 55% of blood is plasma (the fluid portion), and the remaining 45% is composed of solid cells and cell parts. There are three major types of blood cells:

- Red blood cells, or **erythrocytes**, transport oxygen throughout the body. They contain a protein called **hemoglobin** that allows them to carry oxygen. The iron in the hemoglobin gives the cells and the blood their red colors.

- White blood cells, or **leukocytes**, are responsible for fighting infectious diseases and maintaining the immune system. Monocytes, lymphocytes (including B-cells and T-cells), neutrophils, basophils, and eosinophils compose the white blood cells. All are developed in bone marrow. **Monocytes** eat and destroy invaders like bacteria and viruses. **Lymphocytes** are responsible for antibody creation in the defense against invasive organisms and infections. **Neutrophils**, the most abundant white blood cell, take out bacterial and fungal organisms. They are the first line of defense against infections. **Basophils** and mast cells secrete histamine, the substance responsible for itching associated with allergic diseases. **Eosinophils** target parasites and cancer cells, and are part of the body's allergic response. They have low phagocytic activity and primarily secrete destructive enzymes.

- **Platelets** are cell fragments which play a central role in the blood clotting process.

All blood cells in adults are produced in the bone marrow—red blood cells from red marrow and white blood cells from yellow marrow.

Heart

The **heart** is a two-part, muscular pump that forcefully pushes blood throughout the human body. The human heart has four chambers—two upper atria and two lower ventricles separated by a partition called the septum. There is a pair on the left and a pair on the right. Anatomically, *left* and *right* correspond to the sides of the body that the patient themselves would refer to as left and right.

Four valves help to section off the chambers from one another. Between the right atrium and ventricle, the three flaps of the **tricuspid valve** keep blood from backflowing from the ventricle to the atrium, similar to how the two flaps of the **mitral valve** work between the left atrium and ventricle. As these two valves lie between an atrium and a ventricle, they are referred to as **atrioventricular (AV) valves**. The other two valves are **semilunar (SL)** and control blood flow into the two great arteries leaving the ventricles. The **pulmonary valve** connects the right ventricle to the pulmonary artery while the **aortic valve** connects the left ventricle to the aorta.

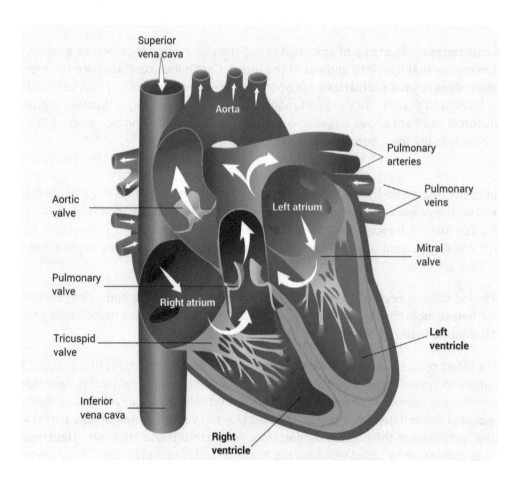

Cardiac Cycle

A **cardiac cycle** is one complete sequence of cardiac activity. The cardiac cycle represents the relaxation and contraction of the heart and can be divided into two phases: diastole and systole.

Diastole is the phase during which the heart relaxes and fills with blood. It gives rise to the diastolic blood pressure (DBP), which is the bottom number of a blood pressure reading. **Systole** is the phase during which the heart contracts and discharges blood. It gives rise to the systolic blood pressure (SBP), which is the top number of a blood pressure reading. The heart's electrical conduction system coordinates the cardiac cycle.

Types of Circulation

Five major blood vessels manage blood flow to and from the heart: the superior and inferior venae cava, the aorta, the pulmonary artery, and the pulmonary vein.

The superior vena cava is a large vein that drains blood from the head and the upper body. The **inferior vena cava** is a large vein that drains blood from the lower body. The **aorta** is the largest artery in the human body and carries blood from the heart to body tissues. The **pulmonary arteries** carry blood from the heart to the lungs. The **pulmonary veins** transport blood from the lungs to the heart.

In the human body, there are two types of circulation: pulmonary circulation and systemic circulation. **Pulmonary circulation** supplies blood to the lungs. Deoxygenated blood enters the right atrium of the heart and is routed through the tricuspid valve into the right ventricle. Deoxygenated blood then travels from the right ventricle of the heart through the pulmonary valve and into the pulmonary arteries. The pulmonary arteries carry the deoxygenated blood to the lungs. In the lungs, oxygen is absorbed, and carbon dioxide is released. The pulmonary veins carry oxygenated blood to the left atrium of the heart.

Systemic circulation supplies blood to all other parts of the body, except the lungs. Oxygenated blood flows from the left atrium of the heart through the mitral, or bicuspid, valve into the left ventricle of the heart. Oxygenated blood is then routed from the left ventricle of the heart through the aortic valve and into the aorta. The aorta delivers blood to the systemic arteries, which supply the body tissues. In the tissues, oxygen and nutrients are exchanged for carbon dioxide and other wastes. The deoxygenated blood along with carbon dioxide and wastes enter the systemic veins, where they are returned to the right atrium of the heart via the superior and inferior vena cava.

Defense Against Infectious Disease

The Immune System

The **immune system** is the body's defense against invading microorganisms (bacteria, viruses, fungi, and parasites) and other harmful, foreign substances. It is capable of limiting or preventing infection.

There are two general types of immunity: innate immunity and acquired immunity. **Innate immunity** uses physical and chemical barriers to block microorganism entry into the body. The biggest barrier is the skin; it forms a physical barrier that blocks microorganisms from entering underlying tissues. Mucous membranes in the digestive, respiratory, and urinary systems secrete mucus to block and remove invading microorganisms. Other natural defenses include saliva, tears, and stomach acids, which are all chemical barriers intended to block infection with microorganisms. Acid is inhospitable to pathogens, as are tears, mucus, and saliva which all contain a natural antibiotic called lysozyme. The respiratory passages contain microscopic cilia which are like bristles that sweep out pathogens. In addition, macrophages and other white blood cells can recognize and eliminate foreign objects through phagocytosis or toxic secretions.

Acquired immunity refers to a specific set of events used by the body to fight a particular infection. Essentially, the body accumulates and stores information about the nature of an invading microorganism. As a result, the body can mount a specific attack that is much more effective than innate immunity. It also provides a way for the body to prevent future infections by the same microorganism.

Acquired immunity is divided into a primary response and a secondary response. The **primary immune response** occurs the first time a particular microorganism enters the body, where macrophages engulf the microorganism and travel to the lymph nodes. In the lymph nodes, macrophages present the invader to helper T lymphocytes, which then activate humoral and cellular immunity. Humoral immunity refers to immunity resulting from antibody production by B lymphocytes. After being activated by helper T lymphocytes, B lymphocytes multiply and divide into plasma cells and memory cells. Plasma cells are B lymphocytes that produce immune proteins called antibodies, or immunoglobulins. Antibodies then bind the microorganism to flag it for destruction by other white blood cells. Cellular immunity refers to the immune response coordinated by T lymphocytes. After being activated by helper T lymphocytes, other T lymphocytes attack and kill cells that cause infection or disease.

The **secondary immune response** takes place during subsequent encounters with a known microorganism. Memory cells respond to the previously encountered microorganism by immediately producing antibodies. Memory cells are B lymphocytes that store information to produce antibodies. The secondary immune response is swift and powerful because it eliminates the need for the time-consuming macrophage activation of the primary immune response. Suppressor T lymphocytes also take part to inhibit the immune response as an overactive immune response could cause damage to healthy cells.

Inflammation occurs if a pathogen evades the barriers and chemical defenses. It stimulates pain receptors, alerting the individual that something is wrong. It also elevates body temperature to speed up chemical reactions, although if a fever goes unchecked it can be dangerous due to the fact that extreme heat unfolds proteins. Histamine is secreted which dilates blood vessels and recruits white blood cells that destroy invaders non-specifically. The immune system is tied to the lymphatic system. The thymus, one of the lymphatic system organs, is the site of maturation of T-cells, a type of white blood cell. The lymphatic system is important in the inflammatory response because lymph vessels deliver leukocytes and collect debris that will be filtered in the lymph nodes and the spleen.

Antigen and Typical Immune Response

Should a pathogen evade barriers and survive through inflammation, an antigen-specific adaptive immune response will begin. Immune cells recognize these foreign particles by their antigens, which are their unique and identifying surface proteins. Drugs, toxins, and transplanted cells can also act as antigens. The body even recognizes its own cells as potential threats in autoimmune diseases.

When a macrophage engulfs a pathogen and presents its antigens, helper T cells recognize the signal and secrete cytokines to signal T lymphocytes and B lymphocytes so that they launch the cell-mediated and humoral response, respectively. The cell-mediated response occurs when the T lymphocytes kill infected cells by secreting cytotoxins. The humoral response occurs when B lymphocytes proliferate into plasma and memory cells. The plasma cells secrete antigen-specific antibodies which bind to the pathogens so that they cannot bind to host cells. Macrophages and other phagocytic cells called

neutrophils engulf and degrade the antibody/pathogen complex. The memory cells remain in circulation and initiate a secondary immune response should the pathogen dare enter the host again.

Active and Passive Immunity

Acquired immunity occurs after the first antigen encounter. The first time the body mounts this immune response is called the primary immune response. Because the memory B cells store information about the antigen's structure, any subsequent immune response causes a secondary immune response which is much faster, and substantially more antibodies are produced due to the presence of memory B cells. If the secondary immune response is strong and fast enough, it will fight off the pathogen before an individual becomes symptomatic. This is a natural means of acquiring immunity.

Vaccination is the process of inducing immunity. **Active immunization** refers to immunity gained by exposure to infectious microorganisms or viruses and can be natural or artificial. **Natural immunization** refers to an individual being exposed to an infectious organism as a part of daily life. For example, it was once common for parents to expose their children to childhood diseases such as measles or chicken pox. **Artificial immunization** refers to therapeutic exposure to an infectious organism as a way of protecting an individual from disease. Today, the medical community relies on artificial immunization as a way to induce immunity.

Vaccines are used for the development of active immunity. A vaccine contains a killed, weakened, or inactivated microorganism or virus that is administered through injection, by mouth, or by aerosol. Vaccinations are administered to prevent an infectious disease but do not always guarantee immunity. Due to circulating memory B cells after administration, the secondary response will fight off the pathogen should it be encountered again in many cases. Both illnesses and vaccinations cause active immunity.

Passive immunity refers to immunity gained by the introduction of antibodies. This introduction can also be natural or artificial. The process occurs when antibodies from the mother's bloodstream are passed on to the bloodstream of the developing fetus. Breast milk can also transmit antibodies to a baby. Babies are born with passive immunity, which provides protection against general infection for approximately the first six months of its life.

Types of Leukocytes

There are many **leukocytes**, or white blood cells, involved in both innate and adaptive immunity. All are developed in bone marrow. Many have been mentioned in the text above, but a comprehensive list is included here for reference.

- Monocytes are large phagocytic cells.
  - Macrophages engulf pathogens and present their antigen. Some circulate, but others reside in lymphatic organs like the spleen and lymph nodes.

  - Dendritic cells are also phagocytic and antigen-presenting.

- Granulocytes are cells that contain secretory granules.
  - Neutrophils are the most abundant white blood cell. They are circulating and aggressive phagocytic cells that are part of innate immunity. They also secrete substances that are toxic to pathogens.

  - Basophils and mast cells secrete histamine which stimulates the inflammatory response.

  - Eosinophils are found underneath mucous membranes and defend against multi-cellular parasites like worms. They have low phagocytic activity and primarily secrete destructive enzymes.

- T lymphocytes mature in the thymus.
  - Helper T cells recognize antigens presented by macrophages and dendritic cells and secrete cytokines that mount the humoral and cell-mediated immune response.

  - Killer T cells are cytotoxic cells involved in the cell-mediated response by recognizing and poisoning infected cells.

  - Suppressor T cells suppress the adaptive immune response when there is no threat to conserve resources and energy.

  - Memory T cells remain in circulation to aid in the secondary immune response.

- B lymphocytes mature in bone marrow.
  - Plasma B cells secrete antigen-specific antibodies when signaled by Helper T cells and are degraded after the immune response.

  - Memory B cells store antigen-specific antibody making instructions and remain circulating after the immune response is over.

- Natural killer cells are part of innate immunity and patrol and identify suspect-material. They respond by secreting cytotoxic substances.

Gas Exchange

Respiratory System

The **respiratory system** enables breathing and supports the energy-making process in our cells. The respiratory system transports an essential reactant, oxygen, to cells so that they can produce energy in their mitochondria via cellular respiration. The respiratory system also removes carbon dioxide, a waste product of cellular respiration.

This system is divided into the upper respiratory system and the lower respiratory system. The **upper system** comprises the nose, the nasal cavity and sinuses, and the pharynx. The **lower respiratory system** comprises the larynx (voice box), the trachea (windpipe), the small passageways leading to the lungs, and the lungs.

The pathway of oxygen to the bloodstream begins with the nose and the mouth. Upon inhalation, air enters the nose and mouth and passes into the sinuses where it gets warmed, filtered, and humidified. The throat, or the pharynx, allows the entry of both food and air; however, only air moves into the

trachea, or windpipe, since the epiglottis covers the trachea during swallowing and prevents food from entering. The trachea contains mucus and cilia. The mucus traps many airborne pathogens while the cilia act as bristles that sweep the pathogens away toward the top of the trachea where they are either swallowed or coughed out.

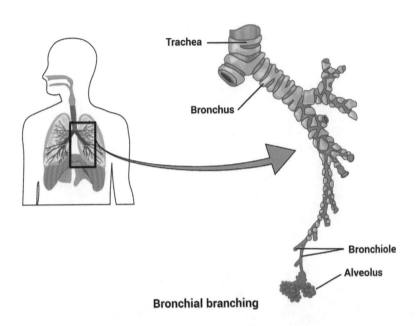

Bronchial branching

The **trachea** itself has two vocal cords at the top that make up the **larynx**. At its bottom, the trachea forks into two major **bronchi**—one for each lung. These bronchi continue to branch into smaller and smaller **bronchioles** before terminating in grape-like air sacs called **alveoli**; these alveoli are surrounded by capillaries and provide the body with an enormous amount of surface area to exchange oxygen and carbon dioxide gases, in a process called **external respiration**.

In total, the lungs contain about 1500 miles of airway passages. The right lung is divided into three lobes (superior, middle, and inferior), and the left lung is divided into two lobes (superior and inferior).

The left lung is smaller than the right lung, likely because it shares its space in the chest cavity with the heart.

A flat muscle underneath the lungs called the **diaphragm** controls breathing. When the diaphragm contracts, the volume of the chest cavity increases and indirectly decreases its air pressure. This decrease in air pressure creates a vacuum, and the lungs pull in air to fill the space. This difference in air pressure that pulls the air from outside of the body into the lungs is a process called negative pressure breathing.

Upon **inhalation** or **inspiration**, oxygen in the alveoli diffuses into the capillaries to be carried by blood to cells throughout the body, in a process called **internal respiration**. A protein called hemoglobin in red blood cells easily bonds with oxygen, removing it from the blood and allowing more oxygen to diffuse in. This protein allows the blood to take in 60 times more oxygen than the body could without it, and this explains how oxygen can become so concentrated in blood even though it is only 21% of the atmosphere. While oxygen diffuses from the alveoli into the capillaries, carbon dioxide diffuses from the

capillaries into the alveoli. When the diaphragm relaxes, the elastic lungs snap back to their original shape; this decreases the volume of the chest cavity and increases the air pressure until it is back to normal. This increased air pressure pushes the carbon dioxide waste from the alveoli through **exhalation** or **expiration**.

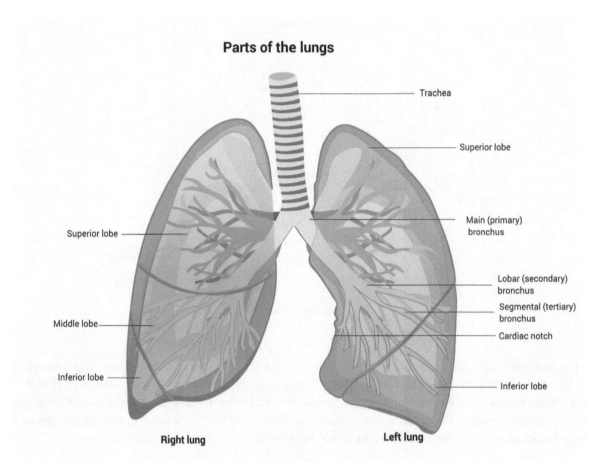

Parts of the lungs

The autonomic nervous system controls breathing. The medulla oblongata gets feedback regarding the carbon dioxide levels in the blood and will send a message to the diaphragm that it is time for a contraction. While breathing can be voluntary, it is mostly under autonomic control.

Functions of the Respiratory System

The respiratory system has many functions. Most importantly, it provides a large area for gas exchange between the air and the circulating blood. It protects the delicate respiratory surfaces from environmental variations and defends them against pathogens. It is responsible for producing the sounds that the body makes for speaking and singing, as well as for non-verbal communication. It also helps regulate blood volume and blood pressure by releasing vasopressin, and it is a regulator of blood pH due to its control over carbon dioxide release, as the aqueous form of carbon dioxide is the chief buffering agent in blood. Erythrocytes use carbonic anhydrase to convert most carbon dioxide in the blood to bicarbonate ions.

Neurons and Synapses

Nervous System

The human **nervous system** coordinates the body's response to stimuli from inside and outside the body. There are two major types of nervous system cells: neurons and neuroglia. **Neurons** are the workhorses of the nervous system and form a complex communication network that transmits electrical impulses termed **action potentials**, while **neuroglia** connect and support them. Motor neurons use sodium and potassium pumps and channels in order to make action potentials occur.

Although some neurons monitor the senses, some control muscles, and some connect the brain to others, all neurons have four common characteristics:

- **Dendrites:** These receive electrical signals from other neurons across small gaps called *synapses*.
- **Nerve cell body:** This is the hub of processing and protein manufacture for the neuron.
- **Axon:** This transmits the signal from the cell body to other neurons.
- **Terminals:** These bridge the neuron to dendrites of other neurons and deliver the signal via chemical messengers called **neurotransmitters.**

Here is an illustration of this:

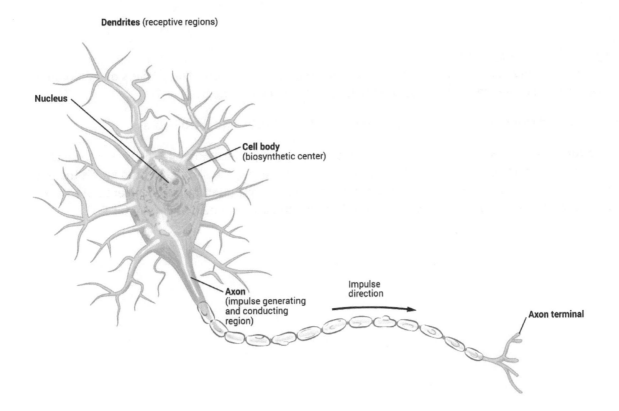

There are two major divisions of the nervous system, central and peripheral:

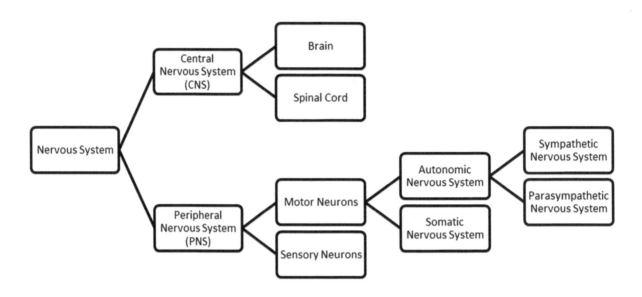

Central Nervous System

The **central nervous system (CNS)** consists of the brain and spinal cord. Three layers of membranes called the meninges cover and separate the CNS from the rest of the body.

The major divisions of the brain are the forebrain, the midbrain, and the hindbrain.

The **forebrain** consists of the cerebrum, the thalamus and hypothalamus, and the rest of the limbic system. The **cerebrum** is the largest part of the brain, and its most well-documented part is the outer cerebral cortex. The cerebrum is divided into right and left hemispheres, and each cerebral cortex hemisphere has four discrete areas, or lobes: frontal, temporal, parietal, and occipital.

The **frontal lobe** governs duties such as voluntary movement, judgment, problem solving, and planning, while the other lobes are more sensory. The **temporal lobe** integrates hearing and language comprehension, the **parietal lobe** processes sensory input from the skin, and the **occipital lobe** functions to process visual input from the eyes. For completeness, the other two senses, smell and taste, are processed via the olfactory bulbs. The thalamus helps organize and coordinate all of this sensory input in a meaningful way for the brain to interpret.

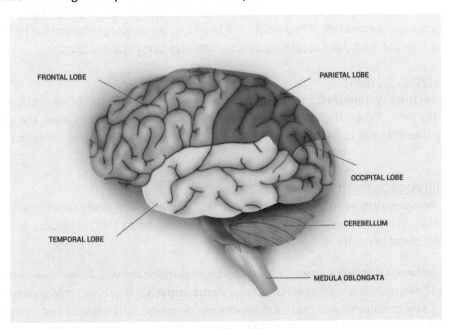

The **hypothalamus** controls the endocrine system and all of the hormones that govern long-term effects on the body. Each hemisphere of the limbic system includes a **hippocampus** (which plays a vital role in memory), an **amygdala** (which is involved with emotional responses like fear and anger), and other small bodies and nuclei associated with memory and pleasure.

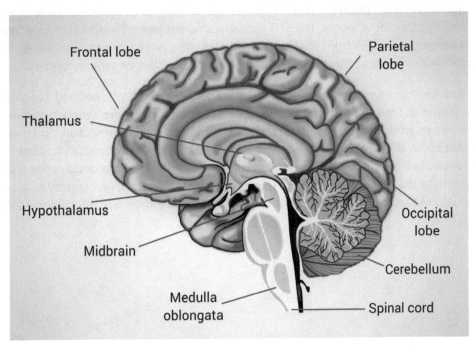

The **midbrain** is in charge of alertness, sleep/wake cycles, and temperature regulation, and it includes the **substantia nigra** which produces melatonin to regulate sleep patterns. The notable components of the **hindbrain** include the **medulla oblongata** and **cerebellum**. The medulla oblongata is located just above the spinal cord and is responsible for crucial involuntary functions such as breathing, heart rate, swallowing, and the regulation of blood pressure. Together with other parts of the hindbrain, the midbrain and medulla oblongata form the **brain stem**. The brain stem connects the spinal cord to the rest of the brain. To the rear of the brain stem sits the cerebellum, which plays key roles in posture, balance, and muscular coordination. The spinal cord itself carries sensory information to the brain and motor information to the body, encapsulated by its protective bony spinal column.

Peripheral Nervous System

The **peripheral nervous system (PNS)** includes all nervous tissue besides the brain and spinal cord. The PNS consists of the sets of cranial and spinal nerves and relays information between the CNS and the rest of the body. The PNS has two divisions: the autonomic nervous system and the somatic nervous system.

Autonomic Nervous System

The **autonomic nervous system (ANS)** governs involuntary, or reflexive, body functions. Ultimately, the autonomic nervous system controls functions such as breathing, heart rate, digestion, body temperature, and blood pressure.

The ANS is split between parasympathetic nerves and sympathetic nerves. These two nerve types are antagonistic and have opposite effects on the body. **Parasympathetic** nerves typically are useful when resting or during safe conditions and decrease heart rate, decrease inhalation speed, prepare digestion, and allow urination and excretion. **Sympathetic** nerves, on the other hand, become active when a person is under stress or excited, and they increase heart rate, increase breathing rates, and inhibit digestion, urination, and excretion.

Somatic Nervous System and the Reflex Arc

The **somatic nervous system (SNS)** governs the conscious, or voluntary, control of skeletal muscles and their corresponding body movements. The SNS contains afferent and efferent neurons. **Afferent** neurons carry sensory messages from the skeletal muscles, skin, or sensory organs to the CNS. **Efferent** neurons relay motor messages from the CNS to skeletal muscles, skin, or sensory organs.

The SNS also has a role in involuntary movements called **reflexes**. A reflex is defined as an involuntary response to a stimulus. They are transmitted via what is termed a **reflex arc**, where a stimulus is sensed by an affector and its afferent neuron, interpreted and rerouted by an interneuron, and delivered to effector muscles by an efferent neuron where they respond to the initial stimulus. A reflex is able to bypass the brain by being rerouted through the spinal cord; the interneuron decides the proper course of action rather than the brain. The reflex arc results in an instantaneous, involuntary response. For example, a physician tapping on the knee produces an involuntary knee jerk referred to as the patellar tendon reflex.

Hormones, Homeostasis, and Reproduction

Endocrine System

The **endocrine system** is made up of the ductless tissues and glands that secrete hormones directly into the bloodstream. It is similar to the nervous system in that it controls various functions of the body, but it does so via secretion of hormones in the bloodstream as opposed to nerve impulses. The endocrine system is also different because its effects last longer than that of the nervous system. Nerve impulses are immediate while hormone responses can last for minutes or even days.

The endocrine system works closely with the nervous system to regulate the physiological activities of the other systems of the body in order to maintain homeostasis. Hormone secretions are controlled by tight feedback loops that are generally regulated by the hypothalamus, the bridge between the nervous and endocrine systems. The hypothalamus receives sensory input via the nervous system and responds by stimulating or inhibiting the pituitary gland which stimulates or inhibits several other glands. The tight control is due to hormone secretions.

Hormones are chemicals that bind to specific target cells. Each hormone will only bind to a target cell that has a specific receptor that has the correct shape. For example, testosterone will not attach to skin cells because skin cells have no receptor that recognizes testosterone.

There are two types of hormones: steroid and protein. **Steroid hormones** are lipid, nonpolar substances, and most are able to diffuse across cell membranes. Once they do, they bind to a receptor that initiates a signal transduction cascade that affects gene expression. Non-steroid hormones bind to receptors on cell membranes that also initiate a signal transduction cascade that affects enzyme activity and chemical reactions.

Major Endocrine Glands

Hypothalamus: This gland is a part of the brain. It connects the nervous system to the endocrine system because it receives sensory information through nerves, and it sends instructions via hormones delivered to the pituitary.

Pituitary Gland: This gland is pea-sized and is found at the bottom of the hypothalamus. It has two lobes called the anterior and posterior lobes. It plays an important role in regulating other endocrine glands. For example, it secretes growth hormone which regulates growth. Other hormones that are released by this gland control the reproductive system, childbirth, nursing, blood osmolarity, and metabolism.

The hormones and glands respond to each other via feedback loops, and a typical feedback loop is illustrated in the picture below. The hypothalamus and pituitary gland are master controllers of most of the other glands.

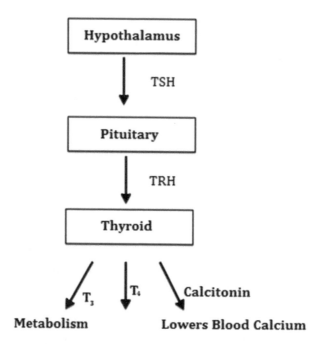

Thymus Gland: This gland is located in the chest cavity, embedded in connective tissue. It produces several hormones that are important for development and maintenance of T lymphocytes, which are important cells for immunity.

Adrenal Gland: One adrenal gland is attached to the top of each kidney. It produces epinephrine and norepinephrine which cause the "fight or flight" response in the face of danger or stress. These hormones raise heart rate, blood pressure, dilate bronchioles, and deliver blood to the muscles. All of these actions increase circulation and release glucose so that the body has an energy burst.

Pineal Gland: The pineal gland secretes melatonin, which is a hormone that regulates the body's circadian rhythms which are the natural wake-sleep cycles.

Testes and Ovaries: They secrete testosterone and both estrogen and progesterone, respectively. They are responsible for secondary sex characteristics, gamete development, and female hormones are important for embryonic development.

Thyroid Gland: This gland releases hormones like thyroxine and calcitonin. Thyroxine stimulates metabolism, and calcitonin monitors the amount of circulating calcium. Calcitonin signals the body to regulating calcium from bone reserves as well as kidney reabsorption of calcium.

Parathyroid Glands: These are four pea-sized glands located on the posterior surface of the thyroid. The main hormone that is secreted is called parathyroid hormone (PTH) which influences calcium levels like calcitonin, except it is antagonistic. PTH increases extracellular levels of calcium while calcitonin decreases it.

Pancreas: The pancreas is an organ that has both endocrine and exocrine functions. It functions outside of a typical feedback loop in that blood sugar seems to signal the pancreas itself. The endocrine functions are controlled by the pancreatic **islets of Langerhans**, which are groups of beta cells scattered throughout the gland that secrete insulin to lower blood sugar levels in the body. Neighboring alpha cells secrete glucagon to raise blood sugar. These complementary hormones keep blood sugar in check.

Reproductive System

The **reproductive system** is responsible for producing, storing, nourishing, and transporting functional reproductive cells, or gametes, in the human body. It includes the reproductive organs, also known as **gonads**, the reproductive tract, the accessory glands and organs that secrete fluids into the reproductive tract, and the **perineal structures**, which are the external genitalia.

Reproduction involves the passing of genes from one generation to the next, and that is accomplished through haploid gametes. Gametes have gone through meiosis and have 23 chromosomes, half the normal number. The male gamete is **sperm**, and the female gamete is an **egg** or **ovum**. When a sperm fertilizes an egg, they create a **zygote**, which is the first cell of a new organism. The zygote has a full set of 46 chromosomes because it received 23 from each parent. Because of sperm and egg development gene shuffling, sperm and egg chromosome sets are all different which results in the variety seen in humans.

Male Reproductive System

The entire **male reproductive system** is designed to generate sperm and produce semen that facilitate fertilization of eggs, the female gametes. The testes are the endocrine glands that secrete **testosterone**, a hormone that is important for secondary sex characteristics and sperm development, or **spermatogenesis**. Testosterone is in the androgen steroid-hormone family. The testes also produce and store 500 million spermatocytes, which are the male gametes, each day. Testes are housed in the **scrotum**, which is a sac that hangs outside the body so that spermatogenesis occurs at cooler and optimal conditions.

The **seminiferous tubules** within the testes produce sperm and then they travel to **epididymis** where they are stored as they mature. Then, the sperm move to the ejaculatory duct via the vas deferens. The ejaculatory duct contains more than just sperm. The **seminal vesicles** secrete an alkaline substance that will help sperm survive in the acidic vagina. The prostate gland secretes enzymes bathed in a milky white fluid that is important for thinning semen after ejaculation to increase its likelihood of reaching the egg. The **bulbourethral**, or Cowper's gland secretes an alkaline fluid that lubricates the urethra prior to ejaculation to neutralize any acidic urine residue.

The sperm, along with all the exocrine secretions, are collectively called **semen**. Their destination is the vagina, and they can only get there if the penis is erect due to arousal and increased circulation. During sexual intercourse, ejaculation will forcefully expel the contents of the semen and effectively deliver the sperm to the egg. The muscular prostate gland is important for ejaculation. Each ejaculation releases 2 to 6 million sperm. Sperm has a whip-like flagellum tail that facilitates movement.

Female Reproductive System

The **vagina** is the passageway that sperm must travel through to reach an egg, the female gamete. Surrounding the vagina are the labia minor and labia major, both of which are folds that protect the urethra and the vaginal opening. The **clitoris** is rich in nerve-endings, making it sensitive and highly stimulated during sexual intercourse. It is above the vagina and urethra. An exocrine gland called the **Bartholin's glands** secretes a fluid during arousal that is important for lubrication.

The female gonads are the ovaries. **Ovaries** generally produce one immature gamete, an egg or oocyte, per month. They are also responsible for secreting the hormones estrogen and progesterone. Fertilization cannot happen unless the ejaculated sperm finds the egg, which are only available at certain times of the month. Eggs, or ova, develop in the ovaries in clusters surrounded by follicles, and after puberty, they are delivered to the uterus once a month via the **Fallopian tubes**. The 28-day average journey of the egg to the uterus is called the menstrual cycle, and it is highly regulated by the endocrine system. The regulatory hormones Gonadotropin releasing hormone (GnRH), luteinizing hormone (LH), and follicle-stimulating hormone (FSH) orchestrate the menstrual cycle. Ovarian hormones estrogen and progesterone are also important in timing as well as for vascularization of the uterus in preparation for pregnancy. **Fertilization** usually happens around ovulation, which is when the egg is inside the fallopian tube. The resulting zygote travels down the tube and implants into the uterine wall. The uterus protects and nourishes the developing embryo for nine months until it is ready for the outside environment.

If the egg released is unfertilized, the uterine lining will slough off during **menstruation**. Should a fertilized egg, called a **zygote**, reach the uterus, it will embed itself into the uterine wall due to uterine vascularization that will deliver blood, nutrients, and antibodies to the developing embryo. The uterus is where the embryo will develop for the next nine months. **Mammary glands** are important female reproductive structures because they produce the milk provided for babies during lactation. Milk contains nutrients and antibodies that benefit the baby.

Practice Questions

1. Which of the following is NOT a major function of the respiratory system in humans?
    a. It provides a large surface area for gas exchange of oxygen and carbon dioxide.
    b. It helps regulate the blood's pH.
    c. It helps cushion the heart against jarring motions.
    d. It is responsible for vocalization.

2. Which of the following areas of the body has the most sweat glands?
    a. Upper back
    b. Arms
    c. Feet
    d. Palms

3. The epidermis is composed of what type of cells?
    a. Osteoclasts
    b. Connective
    c. Dendritic
    d. Epithelial

4. Which of the following functions corresponds to the parasympathetic nervous system?
    a. It stimulates the fight-or-flight response.
    b. It increases heart rate.
    c. It stimulates digestion.
    d. It increases bronchiole dilation.

5. Which of the following is the gland that helps regulate calcium levels?
    a. Osteoid gland
    b. Pineal gland
    c. Parathyroid glands
    d. Thymus gland

Answer Explanations

1. C: Although the lungs may provide some cushioning for the heart when the body is violently struck, this is not a major function of the respiratory system. Its most notable function is that of gas exchange for oxygen and carbon dioxide, but it also plays a vital role in the regulation of blood pH. The aqueous form of carbon dioxide, carbonic acid, is a major pH buffer of the blood, and the respiratory system directly controls how much carbon dioxide stays and is released from the blood through respiration. The respiratory system also enables vocalization and forms the basis for the mode of speech and language used by most humans.

2. A: The upper back has one of the highest densities of sweat glands of any area on the body. While palms, arms, and feet are often thought of as sweaty areas, they have relatively low amounts of sweat glands compared to other parts of the body. Remember that one of the purposes of sweat is thermoregulation, or controlling the temperature of the body. Regulating the temperature of one's core is more important than adjusting the temperature of one's extremities.

3. D: The outermost layer of the skin, the epidermis, consists of epithelial cells. This layer of skin is dead, as it has no blood vessels. Osteoclasts are cells that make up bones. Notice the prefix *Osteo-* which means bone. Connective tissue macrophage cells can be found in a variety of places, and dendritic cells are part of the lymphatic system.

4. C: The parasympathetic nervous system is related to calm, peaceful times without stress that require no immediate decisions. It relaxes the fight-or-flight response, slows heart rate to a comfortable pace, and decreases bronchiole dilation to a normal size. The sympathetic nervous system, on the other hand, is in charge of the fight-or-flight response and works to increase blood pressure and oxygen absorption.

5. C: The parathyroid gland impacts calcium levels by secreting parathyroid hormone (PTH). Osteoid gland is not a real gland. The pineal gland regulates sleep by secreting melatonin, and the thymus gland focuses on immunity. *Thyroid* would also be a correct answer choice as it influences the levels of circulating calcium.

Practice Test

1. Which is true regarding DNA?
   a. It is the genetic code.
   b. It provides energy.
   c. It is single-stranded.
   d. All of the above.

2. Which is the cellular organelle used to tag, package, and ship out proteins destined for other cells or locations?
   a. The Golgi apparatus
   b. The lysosome
   c. The centrioles
   d. The mitochondria

3. How do cellulose and starch differ?
   a. Cellulose and starch are proteins with different R groups.
   b. Cellulose is a polysaccharide made up of glucose molecules and starch is a polysaccharide made up of galactose molecules.
   c. Cellulose and starch are both polysaccharides made up of glucose molecules, but they are connected with different types of bonds.
   d. Cellulose and starch are the same molecule, but cellulose is made by plants and starch is made by animals.

4. Esther is left-handed. Hand dominance is a genetic factor. If being right-handed is a dominant trait over being left-handed, which of the following cannot be true about Esther's parents?
   a. Her parents are both right-handed.
   b. Her parents are both left-handed.
   c. Only one parent is right-handed.
   d. All of the above can be true.

5. What is the probability of AaBbCcDd x AAbbccDD having a child with genotype AAbbccDD?
   a. 1/2
   b. 1/4
   c. 1/8
   d. 1/16

6. Which statement regarding meiosis is correct?
   a. Meiosis produces four diploid cells.
   b. Meiosis contains two cellular divisions separated by interphase II.
   c. Meiosis produces cells with two sets of chromosomes.
   d. Crossing over occurs in the prophase of meiosis I.

7. What molecule serves as the hereditary material for prokaryotic and eukaryotic cells?
   a. Proteins
   b. Carbohydrates
   c. Lipids
   d. DNA

8. What is a product of photosynthesis?
   a. Water
   b. Sunlight
   c. Oxygen
   d. Carbon Dioxide

9. What is cellular respiration?
   a. Making high-energy sugars
   b. Breathing
   c. Breaking down food to release energy
   d. Sweating

10. Which one of the following can perform photosynthesis?
    a. Mold
    b. Ant
    c. Mushroom
    d. Algae

11. Which taxonomic system is commonly used to describe the hierarchy of similar organisms today?
    a. Aristotle system
    b. Linnaean system
    c. Cesalpino system
    d. Darwin system

12. What is the Latin specific name for humans?
    a. Homo sapiens
    b. Homo erectus
    c. Canis familiaris
    d. Homo habilis

13. What organelles have two layers of membranes?
    a. Nucleus, chloroplast, mitochondria
    b. Nucleus, Gogli apparatus, mitochondria
    c. ER, chloroplast, lysosome
    d. Chloroplast, lysosome, ER

14. When human cells divide by meiosis, how many chromosomes do the resulting cells contain?
    a. 96
    b. 54
    c. 46
    d. 23

15. Which is an organelle found in a plant cell but not an animal cell?
    a. Mitochondria
    b. Chloroplast
    c. Golgi body
    d. Nucleus

16. Which of the following correctly expresses the molecular investment for each molecule of G3P produced by the Calvin cycle?
    a. 6 ATP molecules and 9 NADP+ molecules are invested for each molecule of G3P that is produced.
    b. 9 ATP molecules and 6 NADP+ molecules are invested for each molecule of G3P that is produced.
    c. 6 ATP molecules and 9 NADPH molecules are invested for each molecule of G3P that is produced.
    d. 9 ATP molecules and 6 NADPH molecules are invested for each molecule of G3P that is produced.

17. What are the net products of anaerobic glycolysis?
    a. 2 pyruvate molecules, 2 NADH molecules, and 2 ATP molecules
    b. 2 pyruvate molecules, 2 NADH molecules, and 4 ATP molecules
    c. 2 pyruvate molecules, 2 NAD+ molecules, and 2 ATP molecules
    d. 2 pyruvate molecules, 2 NAD+ molecules, and 4 ATP molecules

18. Which of the following correctly states the relationship between the glucose molecules that initially enter into glycolysis and the number of rounds of the citric acid cycle that can be completed?
    a. One round of the citric acid cycle can be completed for each glucose molecule that enters glycolysis.
    b. One round of the citric acid cycle can be completed for every 2 glucose molecules that enter glycolysis.
    c. Two rounds of the citric acid cycle can be completed for each glucose molecule that enters glycolysis.
    d. One round of the citric acid cycle can be completed for every 4 glucose molecule that enter glycolysis.

19. What are the two steps of oxidative phosphorylation?
    a. Electron Transport Chain and Citric Acid Cycle
    b. Krebs Cycle and Electron Transport Chain
    c. Proton Pump and Facilitated Diffusion
    d. Electron Transport Chain and Chemiosmosis

20. Where is the nucleolus located in both plant and animal cells?
    a. Near the chloroplast
    b. Inside the mitochondria
    c. Inside the nucleus
    d. Attached to the cell membrane

21. In which organelle do eukaryotes carry out aerobic respiration?
    a. Golgi apparatus
    b. Nucleus
    c. Mitochondrion
    d. Cytosol

22. What kind of energy do plants use in photosynthesis to create chemical energy?
    a. Light
    b. Electric
    c. Nuclear
    d. Cellular

23. What type of biological molecule is a monosaccharide?
    a. Protein
    b. Carbohydrate
    c. Nucleic acid
    d. Lipid

24. What are the molecular "energy" investments necessary for every G3P molecule produced by the Calvin cycle?
    a. 9 ATP molecules and 6 NADPH molecules
    b. 9 ATP molecules and 6 NADP$^+$ molecules
    c. 6 ATP molecules and 9 NADP$^+$ molecules
    d. 6 ATP molecules and 9 NADPH molecules

25. What does the cell membrane do?
    a. Builds proteins
    b. Breaks down large molecules
    c. Contains the cell's DNA
    d. Controls which molecules are allowed in and out of the cell

26. What structures are made by the body's white blood cells that fight bacterial infections?
    a. Antibodies
    b. Antibiotics
    c. Vaccines
    d. Red blood cells

27. Which blood component is chiefly responsible for oxygen transport?
    a. Platelets
    b. Red blood cells
    c. White blood cells
    d. Plasma cells

28. Where does sperm maturation take place in the male reproductive system?
    a. Seminal vesicles
    b. Prostate gland
    c. Epididymis
    d. Vas Deferens

29. Which hormone in the female reproductive system is responsible for progesterone production?
    a. FSH
    b. LH
    c. hCG
    d. Estrogen

30. Which epithelial tissue comprises the cell layer found in a capillary bed?
    a. Squamous
    b. Cuboidal
    c. Columnar
    d. Stratified

31. Receptors in the dermis help the body maintain homeostasis by relaying messages to what area of the brain?
    a. The cerebellum
    b. The brain stem
    c. The pituitary gland
    d. The hypothalamus

32. When de-oxygenated blood first enters the heart, which of the following choices is in the correct order for its journey to the aorta?
       I. Tricuspid valve → Lungs → Mitral valve
       II. Mitral valve → Lungs → Tricuspid valve
       III. Right ventricle → Lungs → Left atrium
       IV. Left ventricle → Lungs → Right atrium
    a. I and III only
    b. I and IV only
    c. II and III only
    d. II and IV only

33. Which is the simplest nerve pathway that bypasses the brain?
    a. Autonomic
    b. Reflex arc
    c. Somatic
    d. Sympathetic

34. Eosinophils are best described as which of the following?
    a. A type of leukocyte that secretes histamine, which stimulates the inflammatory response.
    b. The most abundant type of leukocyte that secretes substances that are toxic to pathogens.
    c. A type of leukocyte found under mucous membranes that defends against multicellular parasites.
    d. A type of circulating leukocyte that is aggressive and has high phagocytic activity.

35. What types of molecules can move through a cell membrane by passive transport?
    a. Complex sugars
    b. Non-lipid-soluble molecules
    c. Oxygen
    d. Molecules moving from areas of low concentration to areas of high concentration

36. What is ONE feature that both prokaryotes and eukaryotes have in common?
    a. A plasma membrane
    b. A nucleus enclosed by a membrane
    c. Organelles
    d. A nucleoid

37. Diffusion and osmosis are examples of what type of transport mechanism?
    a. Active
    b. Passive
    c. Extracellular
    d. Intracellular

38. How many daughter cells are formed from one parent cell during meiosis?
    a. One
    b. Two
    c. Three
    d. Four

39. The process of breaking large molecules into smaller molecules to provide energy is known as which of the following?
    a. Metabolism
    b. Bioenergetics
    c. Anabolism
    d. Catabolism

40. With which genotype would the recessive phenotype appear, if the dominant allele is marked with "A" and the recessive allele is marked with "a"?
    a. AA
    b. aa
    c. Aa
    d. aA

41. Which organelle is responsible for generating energy for the cell and is referred to as the powerhouse of the cell?
    a. Mitochondria
    b. Nucleus
    c. Ribosomes
    d. Cell wall

42. A color-blind male and a carrier female have three children. What is the probability that they are all color blind?
    a. 1/4
    b. 1/8
    c. 1/16
    d. 1/32

43. What shape does a water molecule form?
    a. C-shape
    b. S-shape
    c. V-shape
    d. T-shape

44. Which type of biological molecule stores information?
    a. Carbohydrates
    b. Nucleic acids
    c. Proteins
    d. Lipids

45. What are chloroplasts responsible for in plant cells?
    a. Maintaining the cell's shape
    b. Containing the cell's DNA
    c. Converting energy from sunlight to glucose
    d. Building proteins

46. What changes in a reaction when an enzyme is added?
    a. The amount of free energy stored in the reactants
    b. The number of products that result from the reaction
    c. The amount of free energy stored in the products
    d. The amount of time needed for the reaction to occur

47. If a molecule were trying to enter an animal cell, which organelle would it have to pass through first?
    a. Cell wall
    b. Cell membrane
    c. Nucleus
    d. Endoplasmic reticulum

48. Which of the following represents a possible number of ATP molecules for oxidative phosphorylation to generate?
    a. 26
    b. 30
    c. 4
    d. 36

49. Which is a product of the photosynthetic reaction?
    a. CO_2
    b. H_2O
    c. O_2
    d. Solar energy

50. Which of the following is identical in both mitosis and meiosis?
    a. The number of divisions
    b. The number of daughter cells produced
    c. The synapsis of homologous chromosomes
    d. When DNA replication occurs

51. Which of the following is directly transcribed from DNA and represents the first step in protein building?
    a. siRNA
    b. rRNA
    c. mRNA
    d. tRNA

52. What information does a genotype give that a phenotype does not?
    a. The genotype necessarily includes the proteins coded for by its alleles.
    b. The genotype will always show an organism's recessive alleles.
    c. The genotype must include the organism's physical characteristics.
    d. The genotype shows what an organism's parents looked like.

	T	t
T		
t		

53. Which statement is supported by the Punnett square above, if "T" = Tall and "t" = short?
    a. Both parents are homozygous tall.
    b. 100% of the offspring will be tall because both parents are tall.
    c. There is a 25% chance that an offspring will be short.
    d. The short allele will soon die out.

54. Which of the following CANNOT be found in a human cell's genes?
    a. Sequences of amino acids to be transcribed into mRNA
    b. Lethal recessive traits like sickle cell anemia
    c. Mutated DNA
    d. DNA that codes for proteins the cell doesn't use

55. Which of Mendel's laws of inheritance states that a dominant phenotype will always be seen over a recessive phenotype?
    a. The law of dominance
    b. The law of similarity
    c. The law of segregation
    d. The law of independent assortment

56. Which base pairs up with adenine in DNA?
    a. Uracil
    b. Thymine
    c. Guanine
    d. Cytosine

57. Which of the following is a special property of water?
    a. Water easily flows through phospholipid bilayers.
    b. A water molecule's oxygen atom allows fish to breathe.
    c. Water is highly cohesive which explains its high melting point.
    d. Water can self-hydrolyze and decompose into hydrogen and oxygen.

58. Which are neurons that transmit signals from the CNS to effector tissues and organs?
    a. Motor
    b. Sensory
    c. Interneuron
    d. Reflex

59. Which statement about white blood cells is true?
    a. B cells are responsible for antibody production.
    b. White blood cells are made in the white/yellow cartilage before they enter the bloodstream.
    c. Platelets, a special class of white blood cell, function to clot blood and stop bleeding.
    d. The majority of white blood cells only activate during the age of puberty, which explains why children and the elderly are particularly susceptible to disease.

60. Which is NOT a function of the pancreas?
    a. Secretes the hormone insulin in response to growth hormone stimulation
    b. Secretes bicarbonate into the small intestine to raise the pH from stomach secretions
    c. Secretes enzymes used by the small intestine to digest fats, sugars, and proteins
    d. Secretes hormones from its endocrine portion in order to regulate blood sugar levels

61. Which action is unrelated to blood pH?
    a. Exhalation of carbon dioxide
    b. Kidney reabsorption of bicarbonate
    c. ADH secretion
    d. Nephron secretion of ammonia

62. Which gland regulates calcium levels?
    a. Thyroid
    b. Pineal
    c. Adrenal
    d. Parathyroid

63. What are the functions of the hypothalamus?
        I. Regulate body temperature
        II. Send stimulatory and inhibitory instructions to the pituitary gland
        III. Receives sensory information from the brain
    a. I and II
    b. I and III
    c. II and III
    d. I, II, and III

64. Which of the following is NOT a component of a sarcomere?
    a. Actin
    b. D-line
    c. B-Band
    d. I-Band

65. Which of the following about the autonomic nervous system (ANS) is true?
    a. It controls the reflex arc
    b. It contains motor (efferent) neurons
    c. It contains sensory (afferent) neurons
    d. It contains both parasympathetic nerves and sympathetic nerves

66. Which of the following systems does not include a transportation system throughout the body?
    a. Cardiovascular system
    b. Endocrine system
    c. Immune system
    d. Nervous system

67. What is the MAIN function of the respiratory system?
    a. To eliminate waste through the kidneys and bladder
    b. To exchange gas between the air and circulating blood
    c. To transform food and liquids into energy
    d. To excrete waste from the body

68. Which is an example of a prezygotic reproductive barrier?
    a. Changing environments
    b. Three species inhabiting the same area
    c. Large feathers
    d. Habitat isolation

69. What type of barrier leads to reproductive isolation after two species mate and produce a hybrid offspring?
    a. Postzygotic barrier
    b. Habitat isolation
    c. Temporal isolation
    d. Behavioral isolation

70. Which concept maintains the variation in a population even as natural selection occurs?
    a. Genetic drift
    b. Sexual selection
    c. Postzygotic barriers
    d. Balancing selection

71. Which idea about the origin of life hypothesizes that microorganisms were transferred to Earth from other objects in the solar system?
    a. Primordial soup
    b. Panspermia
    c. Endosymbiosis
    d. Protocells

72. Protocells are an essential vesicle for replicating what material essential to a hypothesis about the origin of life on Earth?
    a. Clay
    b. RNA
    c. Chromosomes
    d. Carbohydrates

73. What is a driving force behind why speciation can occur?
   a. Geographic separation
   b. Seasons
   c. Daylight
   d. A virus

74. A study that investigated respiration rates of different organisms recorded the data below. Use the data to answer the question below.

	Hours required for 1 g of the animal to use 10 mL O2
Bird	1.3
Human	2.7
Cat	1.9
Elephant	8.3
Lizard	25.2

Which statement best describes the dramatic difference in the data between the lizard and other organisms?
   a. Internal homeostatic mechanisms for temperature regulation are vastly different between lizards and the other animals.
   b. Mitochondria structure is different in endotherms and ectotherms, which results in dramatically different metabolic rates.
   c. The small size of lizards means that it will have the greatest metabolic rate due to its surface area to volume ratio.
   d. Lizards require much less energy to calibrate their temperature because of their habitats.

75. The homology between the circulatory systems of different vertebrates is illustrated in the image below. Which statement best explains the similarities and differences in the physiology between the different organisms?

Fish

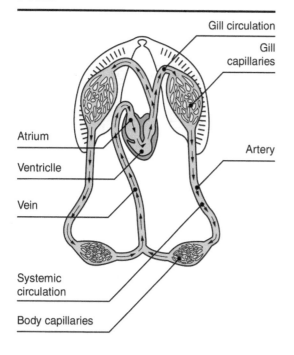

Amphibians

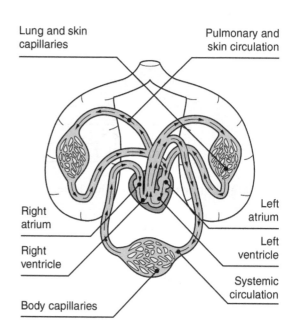

Reptiles

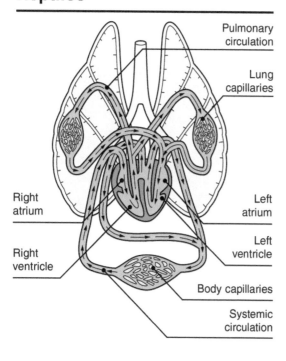

Mammals

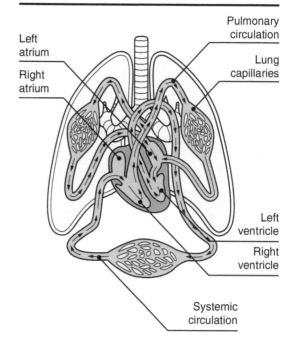

a. The large conservation between amphibians and reptiles suggests that reptiles are the direct descendants of amphibians.

b. Fish are the only group pictured that has gill capillaries, suggesting that there is no common ancestor between fish and the terrestrial groups.

c. The diagrams suggest that reptiles and mammals are more closely related than reptiles and amphibians.

d. The significant difference in the circulatory system between mammals and fish implies that every mammalian organ is more advanced.

76. To answer the following question, refer to the diagram below:

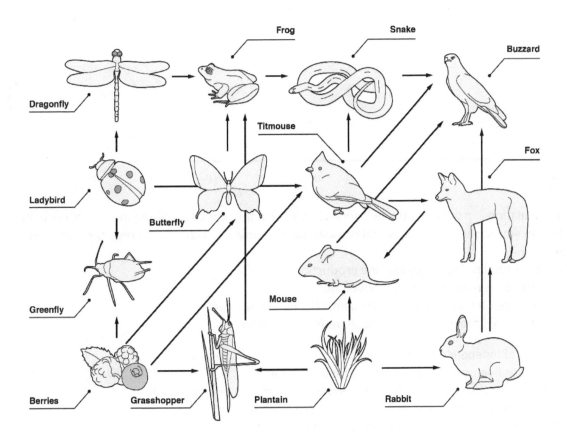

Assume that the snake population has been wiped out by a disease that is only transmutable between snakes. Which of the following would most likely be true as a result?

a. The dragonfly population would decrease.

b. The grasshopper population would increase.

c. The fox population would decrease.

d. The buzzard population would increase.

77. The image below shows hormone levels during a 28-day human menstrual cycle. Days 1–14 are when the follicle develops, and day 14 is ovulation and when the egg is released. During days 15–28, the follicle left behind becomes the corpus luteum. This also is the time when the uterine lining vascularizes and develops.

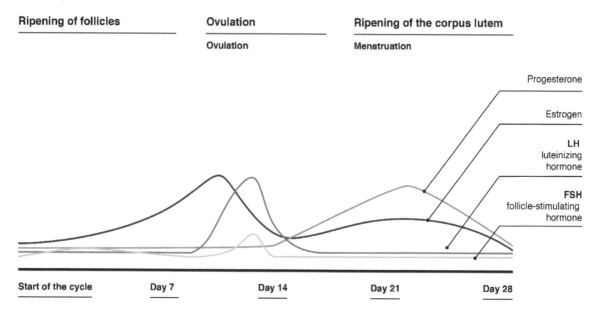

The interrelationship between the hormones is critical in the regulation of the menstrual cycle. Based on the image, which is the most likely and consistently demonstrated conclusion regarding hormone regulation?

    a. FSH is required for progesterone production.

    b. Estrogen is required for LH production.

    c. Progesterone is required for estrogen, LH, and FSH production

    d. FSH and LH require the same negative feedback loop.

78. The law of independent assortment states which of the following?

    a. Tetrads line up in metaphase I in a random fashion

    b. Nondisjunction occurs if there are errors in anaphase I or anaphase II

    c. Sperm will fertilize eggs that have certain characteristics

    d. Dominant alleles mask recessive allele phenotypes

79. Which of the following is *not* true regarding cell cycle checkpoints in mitosis?

    a. A cyclin/CDK pair is responsible for assembling mitotic machinery.

    b. Cyclin protein increases in interphase and is broken down during mitosis.

    c. CDK protein is equally expressed throughout the cell cycle.

    d. G_0 is a state that stimulates progression into S phase.

80. Which statements regarding meiosis are correct?
    I. Meiosis produces four diploid cells.
    II. Meiosis contains two cellular divisions separated by interphase II.
    III. Crossing over occurs in the prophase of meiosis I.

    a. I and II
    b. I and III
    c. II and III
    d. III only

Answer Explanations

1. A: It is the genetic code. Choice *B* is incorrect because DNA does not provide energy—that's the job of carbohydrates and glucose. Choice *C* is incorrect because DNA is double-stranded. Because Choices *B* and *C* are incorrect, Choice *D*, all of the above, is incorrect.

2. A: The Golgi apparatus is designed to tag, package, and ship out proteins destined for other cells or locations. The centrioles typically play a large role only in cell division when they ratchet the chromosomes from the mitotic plate to the poles of the cell. The mitochondria are involved in energy production and are the powerhouses of the cell. The cell structure responsible for cellular storage, digestion, and waste removal is the lysosome. Lysosomes are like recycle bins. They are filled with digestive enzymes that facilitate catabolic reactions to regenerate monomers.

3. C: Cellulose and starch are both polysaccharides, which are long chains of glucose molecules, but they are connected by different types of bonds, which gives them different structures and different functions.

4. D: All of the above. Let's label *R* as the right-handed allele and *r* as the left-handed allele. Esther has to have the combination rr since she's left-handed. She had to get at least one recessive allele from each parent. So, mom could either be Rr or rr (right-handed or left-handed), and dad can also be Rr or rr. As long as each parent carries one recessive allele, it is possible that Esther is left-handed. Therefore, all answer choices are possible.

5. D: 1/16. The probability of each specified genotype can be determined by individual Punnett squares. Each probability should then be then multiplied (law of multiplication) to find the value, which in this case is ½ x ½ x ½ x ½ = 1/16.

	A	a
A	AA	Aa
A	AA	Aa

Probabilities: AA = ½

	B	b
b	Bb	bb
b	Bb	bb

bb = ½

	C	C
c	Cc	cc
c	Cc	cc

cc = ½

	D	d
D	DD	Dd
D	DD	Dd

DD = ½

6. D: Choice *A* is incorrect because meiosis produces haploid cells. Choice *B* is incorrect because there is no interphase II (otherwise gametes would be diploid instead of haploid). Choice *C* is incorrect because the resulting cells only have one set of chromosomes. Choice *D* is the only correct answer because each chromosome set goes through a process called crossing over, which jumbles up the genes on each chromatid, during meiosis I.

7. D: DNA serves as the hereditary material for prokaryotic and eukaryotic cells.

8. C: Oxygen. Water, Choice *A*, is a reactant that gets sucked up by the roots. Carbon dioxide, Choice *D*, is a reactant that goes into the stomata, and sunlight, Choice *B*, inputs energy into the reaction in order to create the high-energy sugar.

9. C: Breaking down food to release energy. Breathing, Choice *B*, is not cellular respiration; breathing is an action that takes place at the organism level with the respiratory system. Making high-energy sugars,

Choice *A,* is photosynthesis, not cellular respiration. Perspiration, Choice *D,* is sweating, and has nothing to do with cellular respiration.

10. D: Algae can perform photosynthesis. One indicator that a plant is able to perform photosynthesis is the color green. Plants with the pigment chlorophyll are able to absorb the warmer colors of the light spectrum, but are unable to absorb green. That's why they appear green. Choices *A* and *C* are types of fungi, and are therefore not able to perform photosynthesis. Fungi obtain energy from food in their environment. Choice *B,* ant, is also unable to perform photosynthesis, since it is an animal.

11. B: The Linnaean system is the commonly used taxonomic system today. It classifies species based on their similarities and moves from comprehensive to more general similarities. The system is based on the following order: species, genus, family, order, class, phylum, and kingdom.

12. A: Homo is the human genus. Sapiens are the only remaining species in the homo genus.

13. A: The nucleus, chloroplast, and mitochondria are all bound by two layers of membrane. The Golgi apparatus, lysosome, and ER only have one membrane layer.

14. D: Human gametes each contain 23 chromosomes. This is referred to as haploid—half the number of the original germ cell (46). Germ cells are diploid precursors of the haploid egg and sperm. Meiosis has two major phases, each of which is characterized by sub-phases similar to mitosis. In Meiosis I, the DNA of the parent cell is duplicated in interphase, just like in mitosis. Starting with prophase I, things become a little different. Two homologous chromosomes form a tetrad, cross over, and exchange genetic content. Each shuffled chromosome of the tetrad migrates to the cell's poles, and two haploid daughter cells are formed. In Meiosis II, each daughter undergoes another division more similar to mitosis (with the exception of the fact that there is no interphase), resulting in four genetically-different cells, each with only ½ of the chromosomal material of the original germ cell.

15. B: Plants use chloroplasts to turn light energy into glucose. Animal cells do not have this ability. Chloroplasts can be found in the plant cell but not the animal cell.

16. D: Both steps of photosynthesis usually occur during daylight because even though the Calvin cycle is not dependent on light energy, it *is* dependent upon the ATP and NADPH produced by the light reactions. This is because that energy can be invested into bonds to create high-energy sugars. For each G3P molecule produced, the Calvin cycle requires the investment of 9 ATP molecules and 6 NADPH molecules.

17. A: The net products of anaerobic glycolysis from one 6-carbon glucose molecule are two 3-carbon pyruvate molecules; 2 reduced nicotinamide adenine dinucleotide (NADH) molecules, which are created when the electron carrier oxidized nicotinamide adenine dinucleotide (NAD+) peels off 2 electrons and a hydrogen atom; and 2 ATP molecules. Glycolysis requires 2 ATP molecules to drive the process forward, and since the gross end product is 4 ATP molecules, the net is 2 ATP molecules.

18. C: Glycolysis produces 2 pyruvate molecules per glucose molecule that gets broken down because glucose is a 6-carbon sugar and pyruvate is a 3-carbon sugar. Each pyruvate molecule oxidizes into a single acetyl-CoA molecule, which then enters the citric acid cycle. Therefore, 2 citric acid cycles can be completed per glucose molecule that initially entered glycolysis.

19. D: Oxidative phosphorylation includes two steps: the electron transport chain and chemiosmosis. These two processes help generate ATP molecules by transferring 2 electrons and a proton (H^+) from each NADH and FADH$_2$ to channel proteins, pumping the hydrogen ions to the inner-membrane space using energy from the high-energy electrons to create a concentration gradient. In chemiosmosis, ATP synthase uses facilitated diffusion to deliver protons across the concentration gradient from the inner mitochondrial membrane to the matrix.

20. C: The nucleolus is always located inside the nucleus. It contains important hereditary information about the cell that is critical for the reproductive process. Chloroplasts, Choice *A*, are only located in plant cells. It is not found in the mitochondria, Choice *B*, or attached to the cell membrane, Choice *D*.

21. C: The mitochondrion is often called the powerhouse of the cell and is one of the most important structures for maintaining regular cell function. It is where aerobic cellular respiration occurs and where most of the cell's ATP is generated. The number of mitochondria in a cell varies greatly from organism to organism and from cell to cell. Cells that require more energy, like muscle cells, have more mitochondria.

22. A: Photosynthesis is the process of converting light energy into chemical energy, which is then stored in sugar and other organic molecules. The photosynthetic process takes place in the thylakoids inside chloroplast in plants. Chlorophyll is a green pigment that lives in the thylakoid membranes and absorbs photons from light.

23. B: Carbohydrates consist of sugars. The simplest sugar molecule is called a monosaccharide and has the molecular formula of CH_2O, or a multiple of that formula. Monosaccharides are important molecules for cellular respiration. Their carbon skeleton can also be used to rebuild new small molecules. Lipids are fats, proteins are formed via amino acids, and nucleic acid is found in DNA and RNA.

24. A: The Calvin cycle is dependent upon the ATP and NADPH produced by the light reactions. Nine ATP molecules and 6 NADPH molecules are invested into the Calvin cycle for every one molecule of glyceraldehyde 3-phosphate (G3P) produced. In the endergonic reduction reaction, NADPH uses energy from ATP to add a hydrogen to each molecule of 3-phosphoglycerate. This converts the 6 molecules of 3-PGA into the 3-carbon sugar G3P. ATP supplies energy by donating a phosphate group (Pi) (becoming ADP), and NADPH loses a hydrogen to become NADP$^+$.

25. D: The cell membrane surrounds the cell and regulates which molecules can move in and out of the cell. Ribosomes build proteins, Choice *A*. Lysosomes, Choice *B*, break down large molecules. The nucleus, Choice *C*, contains the cell's DNA.

26. A: Antibodies. Antibiotics (*B*) fight bacteria, but the body does not make them naturally. White blood cells, not red blood cells (*D*) are the blood cells produced that fight the bacteria. Vaccines (*C*) are given to create antibodies and prevent future illness.

27. B: Red blood cells are the chief transport vehicle for oxygen. Red blood cells contain hemoglobin, a protein that helps transport oxygen throughout the circulatory system.

28. C: The epididymis stores sperm and is a coiled tube located near the testes. The immature sperm that enters the epididymis from the testes migrates through the 20-foot long epididymis tube in about two weeks, where viable sperm are concentrated at the end. The vas deferens is a tube that transports mature sperm from the epididymis to the urethra. Seminal vesicles are pouches attached that add fructose to the ejaculate to provide energy for sperm. The prostate gland excretes fluid that makes up about a third of semen released during ejaculation. The fluid reduces semen viscosity and contains enzymes that aid in sperm functioning; both effects increase sperm motility and ultimate success.

29. C: The female reproductive system is a symphony of different hormones that work together in order to propagate the species. Below, find the function of each one:

Hormone	Source	Action
GnRH	Hypothalamus	Stimulates anterior pituitary to secrete FSH and LH
FSH	Anterior Pituitary	Stimulates ovaries to develop mature follicles (with ova); follicles produce increasingly high levels of estrogen
LH	Anterior Pituitary	Stimulates the release of the ovum by the follicle; follicle then converted into a corpus luteum that secretes progesterone
Estrogen	Ovary (follicle); placenta	Stimulates repair of endometrium of uterus; negative feedback effect inhibits hypothalamus production of GnRH
Progesterone	Ovary (corpus luteum); placenta	Stimulates thickening of and maintains endometrium; negative feedback inhibits pituitary production of LH
Prolactin	Anterior pituitary	Stimulates milk production after childbirth
Oxytocin	Posterior pituitary	Stimulates milk "letdown"
Androgens	Adrenal glands	Stimulates sexual drive
hCG	Embryo (if pregnancy)	Stimulates production of progesterone

30. A: Epithelial cells line cavities and surfaces of body organs and glands, and the three main shapes are squamous, columnar, and cuboidal. Epithelial cells contain no blood vessels, and their functions involve absorption, protection, transport, secretion, and sensing. Simple squamous epithelial are flat cells that are present in lungs and line the heart and vessels. Their flat shape aids in their function, which is diffusion of materials. The tunica intima, the inner layer of blood vessels, is lined with simple squamous epithelial tissue that sits on the basement membrane. Simple cuboidal epithelium is found in ducts, and simple columnar epithelium is found in tubes with projections (uterus, villi, bronchi). Any of these types of epithelial cells can be stacked, and then they are called stratified, not simple.

31. D: Receptors in the dermis help the body maintain homeostasis, for example, in terms of regulating body temperature. Signals travel to the hypothalamus, which then secretes hormones that activate effectors to keep internal temperature at a set point of 98.6°F (37°C). For example, if the environment is too cold, the hypothalamus will initiate a pathway that will cause the muscles to shiver because shivering helps heat the body.

32. A: Carbon dioxide rich blood is delivered and collected in the right atrium and moved to the right ventricle. The tricuspid valve prevents backflow between the two chambers. From there, the pulmonary artery takes blood to the lungs where diffusion causes gas exchange. Then, blood collects in the left atrium and moves to the left ventricle. The mitral valve prevents the backflow of blood from the ventricle to the atrium. Finally, blood is pumped to the body and released in the aorta.

33. B: The reflex arc is the simplest nerve pathway. The stimulus bypasses the brain, going from sensory receptors through an afferent (incoming) neuron to the spinal cord. It synapses with an efferent (outgoing) neuron in the spinal cord and is transmitted directly to muscle. There is no interneuron involved in a reflex arc. The classic example of a reflex arc is the knee jerk response. Tapping on the patellar tendon of the knee stretches the quadriceps muscle of the thigh, resulting in contraction of the muscle and extension of the knee.

34. C: Eosinophils, like neutrophils, basophils, and mast cells, are a type of leukocyte in a class called granulocytes. They are found underneath mucous membranes in the body and they primarily secrete destructive enzymes and defend against multicellular parasites like worms. Choice *A* describes basophils and mast cells, and Choices *B* and *D* describe neutrophils. Unlike neutrophils, which are aggressive phagocytic cells, eosinophils have low phagocytic activity.

35. C: Molecules that are soluble in lipids, like fats, sterols, and vitamins (A, D, E and K), for example, are able to move in and out of a cell using passive transport. Water and oxygen are also able to move in and out of the cell without the use of cellular energy. Complex sugars and non-lipid soluble molecules are too large to move through the cell membrane without relying on active transport mechanisms. Molecules naturally move from areas of high concentration to those of lower concentration. It requires active transport to move molecules in the opposite direction, as suggested by Choice *D*.

36. A: Both types of cells are enclosed by a cell membrane, which is selectively permeable. Selective permeability means essentially that it is a gatekeeper, allowing certain molecules and ions in and out, and keeping unwanted ones at bay, at least until they are ready for use. Prokaryotes contain a nucleoid and do not have organelles; eukaryotes contain a nucleus enclosed by a membrane, as well as organelles.

37. B: Diffusion and osmosis are examples of passive transport. Unlike active transport, passive transport does not require cellular energy. Diffusion is the movement of particles, such as ions, nutrients, or waste, from high concentration to low. Osmosis is the spontaneous movement of water from an area of high concentration to one of low concentration. Facilitated diffusion is another type of passive transport where particles move from high concentration to low concentration via a protein channel.

38. D: Meiosis has the same phases as mitosis, except that they occur twice—once in meiosis I and once in meiosis II. During meiosis I, the cell splits into two. Each cell contains two sets of chromosomes. Next, during meiosis II, the two intermediate daughter cells divide again, producing four total haploid cells that each contain one set of chromosomes.

39. D: Catabolism is the process of breaking large molecules into smaller molecules to release energy for work. Carbohydrates and fats are catabolized to provide energy for exercise and daily activities. Anabolism synthesizes larger molecules from smaller constituent building blocks. Bioenergetics and metabolism are more general terms involving overall energy production and usage.

40. B: Dominant alleles are considered to have stronger phenotypes and, when mixed with recessive alleles, will mask the recessive trait. The recessive trait would only appear as the phenotype when the allele combination is "aa" because a dominant allele is not present to mask it.

41. A: Mitochondria is described as the powerhouse of the cell. The nucleus, Choice *B*, contains the cell's DNA. The ribosomes, Choice *C*, build proteins. The cell wall, Choice *D*, maintains the shape of plant cells and protects its contents.

42. B: 1/8. Color blindness is a recessive, sex-linked trait. The Punnett square below shows the cross between a carrier female and a color-blind male. The two offspring in bold are color blind. The probability of having one child that is color blind is ½. The probability of having three color-blind children is ½ x ½ x ½ (law of multiplication) or 1/8.

	X^C	X^c
X^c	**$X^C X^c$**	**$X^c X^c$**
Y	$X^C Y$	$X^c Y$

43. C: The best answer is Choice C. Water molecules form a V-shape because of the uneven sharing of electrons between the atoms. The oxygen atom is slightly negatively charged, and the hydrogen atoms are slightly positively charged, so they pull away from each other and the molecule forms a V-shape.

44. B: The correct answer is Choice B. Nucleic acids include DNA and RNA, which store an organism's genetic information. Carbohydrates, Choice A, are used as an energy source for an organism. Proteins, Choice C, are important for the structure and function of organisms. Lipids, Choice D, are important for energy storage and insulation.

45. C: Chloroplasts are responsible for photosynthesis in plant cells, which is the process of converting sunlight energy to glucose energy. The cell wall helps maintain the cell's shape, Choice A. The nucleus contains the cell's DNA, Choice B. Ribosomes build proteins, Choice D.

46. D: The correct answer is Choice D. Enzymes increase the rate of the metabolic reaction. They do not change the starting or ending amount of free energy in the reactants or products, so Choices A and C are incorrect. The number of products also remains the same, so Choice B is incorrect.

47. B: The correct answer is Choice B. An animal cell is surrounded by a cell membrane. The cell membrane contains proteins that regulate which molecules are allowed in and out of the cell. Only plant cells are surrounded by a cell wall, so Choice A is incorrect. The nucleus, Choice C, is located in middle of the cell and would not be the first organelle that a molecule would encounter. The endoplasmic reticulum, Choice D, is located within the cytoplasm, inside the cell membrane.

48. A: The correct answer is Choice A. Oxidative phosphorylation is the final part of aerobic respiration and generates between 26 and 28 ATP molecules. In total, aerobic respiration generates between 30 and 32 ATP molecules, Choice B. Glycolysis and the citric acid cycle are the first parts of aerobic respiration and generate 4 ATP molecules, Choice C.

49. C: The correct answer is Choice C. On the right side of the chemical equation, O_2 is one of the products of the reaction. The left side of the equation contains the reactants. The reactants are CO_2, which is Choice A, and H_2O, which is Choice B. Solar energy, Choice D, joins the reactants to drive the reaction forward.

50. D: The correct answer is Choice D. In both mitosis and meiosis, DNA replication occurs during interphase. Mitosis has one cell division whereas meiosis has two cell divisions; therefore, Choice A, the number of divisions, is incorrect. Mitosis produces two daughter cells whereas meiosis produces four daughter cells; therefore, Choice B is incorrect. Synapsis of homologous chromosomes, Choice C, does not occur in mitosis.

51. C: mRNA is directly transcribed from DNA before being taken to the cytoplasm and translated by rRNA into a protein. tRNA transfers amino acids from the cytoplasm to the rRNA for use in building these proteins. siRNA is a special type of RNA which interferes with other strands of mRNA typically by causing them to get degraded by the cell rather than translated into protein.

52. B: Since the genotype is a depiction of the specific alleles that an organism's genes code for, it includes recessive genes that may or may not be otherwise expressed. The genotype does not have to name the proteins that its alleles code for; indeed, some of them may be unknown. The phenotype is the physical, visual manifestations of a gene, not the genotype. The genotype does not necessarily include any information about the organism's physical characters. Although some information about an organism's parents can be obtained from its genotype, its genotype does not actually show the parents' phenotypes.

53. C: One in four offspring (or 25%) will be short, so all four offspring cannot be tall. Although both of the parents are tall, they are hybrid or heterozygous tall, not homozygous. The mother's phenotype is for tall, not short. A Punnett square cannot determine if a short allele will die out. Although it may seem intuitive that the short allele will be expressed by lower numbers of the population than the tall allele, it still appears in 75% of the offspring (although its effects are masked in 2/3 of those). Besides, conditions could favor the recessive allele and kill off the tall offspring.

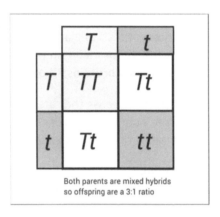

Both parents are mixed hybrids
so offspring are a 3:1 ratio

54. A: Human genes are strictly DNA and do not include proteins or amino acids. A human's genome and collection of genes will include even their recessive traits, mutations, and unused DNA.

55. A: The law of dominance states that dominant traits are always passed on and will be present in future generations, even if recessive alleles are inherited. Choice *B*, the law of similarity, is not one of Mendel's laws. The law of segregation, Choice *C*, states that the alleles of a gene are separated during meiosis. The law of independent assortment, Choice *D*, states that genes on the same chromosome are all independently and randomly sorted and assigned during the second division of meiosis.

56. B: The correct answer is Choice *B*. In DNA, adenine pairs only with thymine. Uracil, Choice *A*, is only found in RNA, not DNA. Guanine and cytosine, Choices *C* and *D*, pair with each other exclusively in DNA.

57. C: Water's polarity lends it to be extremely cohesive and adhesive; this cohesion keeps its atoms very close together. Because of this, it takes a large amount of energy to melt and boil its solid and liquid forms. Phospholipid bilayers are made of nonpolar lipids and water, a polar liquid, cannot flow through it. Cell membranes use proteins called aquaporins to solve this issue and let water flow in and out. Fish breathe by capturing dissolved oxygen through their gills. Water can self-ionize, wherein it decomposes into a hydrogen ion (H^+) and a hydroxide ion (OH^-), but it cannot self-hydrolyze.

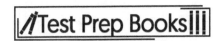

58. A: Motor neurons transmit signals from the CNS to effector tissues and organs, such as skeletal muscle and glands. Sensory neurons carry impulses from receptors in the extremities to the CNS. Interneurons relay impulses from neuron to neuron.

59. A: When activated, B cells create antibodies against specific antigens. White blood cells are generated in yellow bone marrow, not cartilage. Platelets are not a type of white blood cell, and are typically cell fragments produced by megakaryocytes. White blood cells are active throughout nearly all of one's life, and have not been shown to specially activate or deactivate because of life events like puberty or menopause.

60. A: The exocrine portion of the pancreas (the majority of it) is an accessory organ to the digestive system (meaning that food never touches it—it is not part of the alimentary canal). It secretes bicarbonate to neutralize stomach acid and enzymes to aid in digestion. It also regulates blood sugar levels through the complementary action of insulin and glucagon that are located in the Islets of Langerhans (endocrine portion). Choice *A* is incorrect because it is not the growth hormone that stimulates insulin secretion, but rather blood sugar levels.

61. C: ADH secretion is correct. Antidiuretic hormone controls water reabsorption. In its presence, water is reabsorbed, and urine is more concentrated. When absent, water is excreted, and urine is dilute. It is a regulator of blood volume, not pH. The other choices do affect blood pH.

62. D: The gland that regulates blood calcium levels is the parathyroid gland. Humans have four parathyroid glands located by the thyroid on each side of the neck, just below the larynx. Typical with the endocrine system, the parathyroid glands operate via feedback loops. If calcium in the blood is low, the parathyroid glands produce parathyroid hormone, which circulates to the bones and removes calcium. If calcium is high, they turn off parathyroid hormone production.

63. D: The hypothalamus is the link between the nervous and endocrine system. It receives information from the brain and sends signals to the pituitary gland, instructing it to release or inhibit release of hormones. Aside from its endocrine function, it controls body temperature, hunger, sleep, circadian rhythms, and is part of the limbic system.

64. B: The smallest unit of a muscle fiber, sarcomeres, contain the actin and myosin proteins responsible for the mechanical process of muscle contractions. Located between two Z-lines, the actin and myosin filaments are configured in parallel, end-to-end, along the entire length of the myofibril. The sarcomere consists of four segments: the A-band, H-zone, I-band, and Z-line. The B-band and D-line are fictitious and are not components of a sarcomere.

65. D: The autonomic nervous system (ANS), a division of the PNS, controls involuntary functions such as breathing, heart rate, blood pressure, digestion, and body temperature via the antagonistic parasympathetic and sympathetic nerves. Choices *A, B,* and *C* are incorrect because they describe characteristics of the somatic nervous system, which is the other division of the PNS. It controls skeletal muscles via afferent and efferent neurons. Afferent neurons carry sensory messages from skeletal muscles, skin, or sensory organs to the CNS, while Efferent neurons relay motor messages from the CNS to skeletal muscles, skin, or sensory organs. While skeletal muscle movement is under voluntary control, the somatic nervous system also plays a role in the involuntary reflex arc.

66. B: The endocrine system's organs are glands which are spread throughout the body. The endocrine system itself does not connect the organs or transport the hormones they secrete. Rather, the various glands secrete the hormone into the bloodstream and lets the cardiovascular system pump it throughout the body. The other three body systems each include a network throughout the body:

- Cardiovascular system: veins and arteries
- Immune system: lymphatic vessels (it does also use the circulatory system)
- Nervous system: nerve networks

67. B: The respiratory system mediates the exchange of gas between the air and the circulating blood, mainly by the act of breathing. It filters, warms, and humidifies the air that gets breathed in and then passes it into the blood stream. The digestive system transforms food and liquids into energy and helps excrete waste from the body. Eliminating waste via the kidneys and bladder is a function of the urinary system.

68. D: Prezygotic barriers prevent fertilization. These include habitat isolation, temporal isolation, and behavioral isolation. Two species may inhabit the same area but don't often encounter each other, which is habitat isolation.

69. A: Postzygotic barriers contribute to reproductive isolation after fertilization. The mixed species offspring are called hybrids. Once the hybrid zygote is formed, it may have reduced viability. These hybrids may not survive, and if they do, they are very frail. If a hybrid survives, it may have reduced fertility. In some cases, hybrids are actually viable and fertile for the first generation. However, when the hybrids mate with each other or with one of the parent species, subsequent generations are weak or sterile.

70. D: Although the idea of natural selection supports the survival and reproduction of the individuals best suited for their environment, there's always variation within a population. This idea is known as balancing selection, which maintains the variation in a population through the heterozygote advantage and frequency-dependent selection. As the environment is always changing, the characteristics necessary to increase or maintain survival and reproduction of a species may also change.

71. B: Panspermia is the idea that life came to Earth from other areas of the universe. It hypothesizes that meteoroids, asteroids, and other small objects from the solar system landed on Earth and transferred microorganisms to the Earth's surface. This theory proposes that there were seeds of life everywhere in the universe, and when these seeds were brought to Earth, the conditions were ideal for living organisms to develop and flourish.

72. B: Protocells are small, round groups of lipids that are hypothesized to be responsible for the origin of life. They are proposed to have been formed by vesicles, which are fluid-filled compartments enclosed by a membrane-like structure. Vesicles form spontaneously when lipids, such as protocells, are added to water. Some of the clay that covered the Earth is believed to contain RNA, which the vesicles could encapsulate. Vesicles that carried RNA could then divide and have replicated RNA in its daughter cell, increasing the amount of genetic material in the environment. The RNA inside of them may have been used as templates to create more stable DNA strands. Then, according to evolutionary theory, after the formation of DNA, the origin of life and more complex living organisms began.

73. A: Speciation is the method by which one species splits into two or more species. In allopatric speciation, one population is divided into two subpopulations. If a drought occurs and a large lake becomes divided into two smaller lakes, each lake is left with its own population that cannot intermingle

with the population of the other lake. When the genes of these two subpopulations are no longer mixing with each other, new mutations can arise and natural selection can take place.

74. A: The data table is referring to rate of respiration. The more time required to use oxygen indicates a lower rate of respiration. The quicker the animal uses up oxygen, the more aerobic respiration is taking place. Ectothermic and endothermic animals are different, since endotherms have to regulate their own internal temperature, which requires more ATP because it is controlled by an elaborate feedback loop. More ATP requires higher levels of respiration. Choice *C* is attractive because it is true that smaller organisms respire faster due to their larger surface area to volume ratio. However, Choice *C* is incorrect for two reasons. First, not all lizards are smaller than the other organisms. Secondly, the dramatic nature of the difference is due to the lizards' simpler mode of temperature regulation. Choice *D* is attractive because lizards bask in sunny habitats to help maintain ideal temperatures, but they lack a feedback loop that calibrates temperature, so it is incorrect. Choice *B* is incorrect because mitochondrial structure is the same among all eukaryotes—it does not change.

75. C: The conservation of the four-chambered heart provides the justification for concluding that the greater relatedness is between mammals and reptiles. As this is somewhat subjective due to qualitative interpretation, it is still the best choice. Choice *A* is incorrect because organisms are not necessarily direct descendants of more primitive versions of existing organisms; they could have branched from along different points in the phylogenetic tree. Choice *B* is incorrect because all organisms share a common ancestor. Choice *D* is incorrect because it is impossible to extrapolate broad conclusions based on qualitative data regarding one body system. Fish do, in fact, have very complicated organs such as gills that are highly differentiated and complex.

76. A: If the snake population were to disappear, the organisms that it hunted would increase, and the food that those organisms ate would decrease. Without the snake, that means there are more frogs, which in turn means there are more frogs to eat more dragonflies, so the dragonfly population would decrease. Choice *B* is incorrect, since an increase in frogs means more frogs would eat more grasshoppers, so grasshoppers would decrease as opposed to increase. Choice *C* is incorrect because fewer snakes means more titmice, which provide more food for the foxes and results in an increase in the fox population, not a decrease. Choice *D* is incorrect because, even though buzzards have food sources other than snakes, losing one of their food sources would help to decrease the population, not increase it.

77. D: This data suggests that FSH and LH are both regulated by the same mechanism because they spike at the same time. Choice *A* is incorrect because while right after the FSH spike, progesterone levels increase, FSH immediately decreases and progesterone is unaffected. Choice *B* is incorrect because estrogen has no effect on LH in the luteal phase. Choice *C* is not the best choice because in the follicular phase, the levels of progesterone, LH and FSH are all low.

78. A: Tetrads line up in metaphase I in a random fashion. Choice *B* is not a heredity issue; nondisjunction is a mistake in chromatid separation that is not inherited. Choice *C* is untrue because fertilization is random. Choice *D* explains alleles, but does not explain the mechanism behind genetic diversity like Choice *A* does. The law of independent assortment pertains to the random lineup of chromosomes in metaphase I.

79. D: G_0 is a state that stimulates progression into S phase. G_0 is actually a checkpoint that involves cells exiting the cell cycle, such as mature neurons and damaged cells that may undergo apoptosis. Therefore, Choice *D* is the untrue statement. Choices *A*, *B*, and *C* refer to the Maturation Promoting

Factor (MPF), which is a cyclin/CDK pair that forms in G_2 when rising levels of cyclin bind to and activate an ever-present cyclin dependent kinase.

80. D: III only. Roman numeral I is incorrect because meiosis produces haploid cells. Roman numeral II is incorrect because there is no interphase II (otherwise gametes would be diploid instead of haploid). Choice *D* is the only correct answer because the others contain Roman numerals I and II.

Dear IB Biology Test Taker,

We would like to start by thanking you for purchasing this study guide for your IB Biology exam. We hope that we exceeded your expectations.

Our goal in creating this study guide was to cover all of the topics that you will see on the test. We also strove to make our practice questions as similar as possible to what you will encounter on test day. With that being said, if you found something that you feel was not up to your standards, please send us an email and let us know.

We have study guides in a wide variety of fields. If you're interested in one, try searching for it on Amazon or send us an email.

Thanks Again and Happy Testing!
Product Development Team
info@studyguideteam.com

FREE Test Taking Tips DVD Offer

To help us better serve you, we have developed a Test Taking Tips DVD that we would like to give you for FREE. **This DVD covers world-class test taking tips that you can use to be even more successful when you are taking your test.**

All that we ask is that you email us your feedback about your study guide. Please let us know what you thought about it – whether that is good, bad or indifferent.

To get your **FREE Test Taking Tips DVD**, email freedvd@studyguideteam.com with "FREE DVD" in the subject line and the following information in the body of the email:

a. The title of your study guide.

b. Your product rating on a scale of 1-5, with 5 being the highest rating.

c. Your feedback about the study guide. What did you think of it?

d. Your full name and shipping address to send your free DVD.

If you have any questions or concerns, please don't hesitate to contact us at freedvd@studyguideteam.com.

Thanks again!

Made in United States
Troutdale, OR
08/11/2023

11984225R00093